Audio Demystified

Demystified Series

Accounting Demystified
Advanced Statistics Demystified
Algebra Demystified
Alternative Energy Demystified
Anatomy Demystified
asp.net 2.0 Demystified
Astronomy Demystified
Audio Demystified
Biology Demystified
Biotechnology Demystified
Business Calculus Demystified
Business Math Demystified
Business Statistics Demystified
C++ Demystified
Calculus Demystified
Chemistry Demystified
College Algebra Demystified
Corporate Finance Demystified
Databases Demystified
Data Structures Demystified
Differential Equations Demystified
Digital Electronics Demystified
Earth Science Demystified
Electricity Demystified
Electronics Demystified
Environmental Science Demystified
Everyday Math Demystified
Forensics Demystified
Genetics Demystified
Geometry Demystified
Home Networking Demystified
Investing Demystified
Java Demystified
JavaScript Demystified
Linear Algebra Demystified
Macroeconomics Demystified
Management Accounting Demystified
Math Proofs Demystified
Math Word Problems Demystified
Medical Billing and Coding Demystified
Medical Terminology Demystified
Meteorology Demystified
Microbiology Demystified
Microeconomics Demystified
Nanotechnology Demystified
Nurse Management Demystified
OOP Demystified
Options Demystified
Organic Chemistry Demystified
Personal Computing Demystified
Pharmacology Demystified
Physics Demystified
Physiology Demystified
Pre-Algebra Demystified
Precalculus Demystified
Probability Demystified
Project Management Demystified
Psychology Demystified
Quality Management Demystified
Quantum Mechanics Demystified
Relativity Demystified
Robotics Demystified
Signals and Systems Demystified
Six Sigma Demystified
SQL Demystified
Statics and Dynamics Demystified
Statistics Demystified
Technical Math Demystified
Trigonometry Demystified
UML Demystified
Visual Basic 2005 Demystified
Visual C# 2005 Demystified
XML Demystified

Audio Demystified

STAN GIBILISCO

New York Chicago San Francisco Lisbon London Madrid
Mexico City Milan New Delhi San Juan Seoul
Singapore Sydney Toronto

The McGraw-Hill Companies

Cataloging-in-Publication Data is on file with the Library of Congress.

1 2 3 4 5 6 7 8 9 0 DOC/DOC 0 1 0 9 8 7 6

ISBN 13: 978-0-07-146983-8
ISBN 10: 0-07-146983-4

The sponsoring editor for this book was Judy Bass, the editing supervisor was David E. Fogarty, and the production supervisor was Pamela A. Pelton. It was set in Times Roman by D & P Editorial Services, LLC. The art director for the cover was Margaret Webster-Shapiro.

Printed and bound by RR Donnelley.

This book was printed on acid-free paper.

To Tony, Samuel, and Tim

ABOUT THE AUTHOR

Stan Gibilisco is one of McGraw-Hill's most prolific and popular authors. His clear, reader-friendly writing style makes his books accessible to a wide audience, and his experience as an electronics engineer, researcher, and mathematician makes him an ideal editor for reference books and tutorials. Stan has authored several titles for the McGraw-Hill *Demystified* library of home-schooling and self-teaching volumes, along with more than 30 other books and dozens of magazine articles. His work has been published in several languages. *Booklist* named his *McGraw-Hill Encyclopedia of Personal Computing* one of the "Best References of 1996," and named his *Encyclopedia of Electronics* one of the "Best References of the 1980s."

CONTENTS

PREFACE

This book is for people who want to learn basic audio electronics theory without taking a formal course. It can serve as a classroom supplement, a tutorial aid, or a home-schooling text. If you have worked with audio systems but don't know exactly what goes on at the component level, or if you want to improve your understanding of technical theory and jargon, this book should help you. Maybe this book will even stimulate your inventive imagination, although it is not a design guide.

As you take this course, you'll encounter multiple-choice quizzes and a final exam to help you measure your progress. All quiz and exam questions are composed like those in standardized tests. The quizzes are "open-book." You may refer to the chapter texts when taking them. The final exam contains questions drawn uniformly from all the chapters. It is a "closed-book" test. Don't look back at the text when taking it. Answers to all quiz and exam questions are listed in an appendix at the back of the book.

You don't need a high-level math or science background for this course. Middle-school algebra, geometry, and physics will suffice. You'll have an easy time here if you've been through "basic training" in electricity and electronics. In particular, you should be able to read simple circuit diagrams. If you need preparation in these areas, *Electricity Demystified* and *Electronics Demystified* (studied in that order) are recommended as prerequisites. A more comprehensive overall treatment is provided by *Teach Yourself Electricity and Electronics*. All three books are published by McGraw-Hill.

I recommend that you complete one chapter of this book per week. That way, in a few months, you'll finish the course. You can then use this book, with its comprehensive index, as a permanent reference.

Suggestions for future editions are welcome.

Stan Gibilisco

Audio Demystified

CHAPTER 1

Direct Current Basics

Anyone who is serious about audio electronics must be familiar with the fundamentals of direct current (DC) phenomena and systems. In this chapter, you'll learn how current, voltage, resistance, and power are related in DC circuits.

Symbols and Diagrams

In technical diagrams of electrical and electronic circuits, the simplest symbol is the one representing a wire or *electrical conductor*: a straight, solid line. Sometimes, dashed lines are used to represent conductors, but usually, dashed lines are drawn to partition diagrams into constituent circuits, or to indicate that certain components interact with each other or operate in step with each other. Conductor lines are almost always drawn either horizontally across, or vertically up and down the page. This keeps the diagram neat and easy to read.

When two conductor lines cross in a diagram, they are *not* meant to be connected at the crossing point unless a heavy, black dot is placed there. The dot

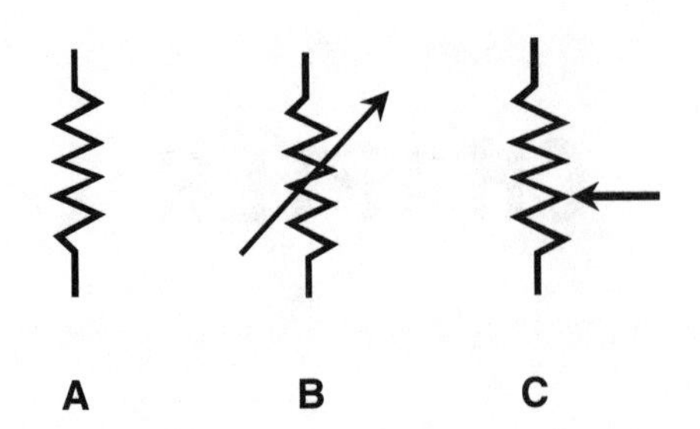

Figure 1-1 Schematic symbols for a fixed resistor (A), a two-terminal variable resistor (B), and a three-terminal potentiometer (C).

should always be clearly visible wherever conductors are to be connected, no matter how many conductors meet at the junction. A *resistor* is indicated by a zig-zag. A variable resistor, or *potentiometer*, is indicated by a zig-zag with an arrow through it, or by a zig-zag with an arrow pointing at it. These symbols are shown in Fig. 1-1.

An *electrochemical cell* (such as a common flashlight "battery") is shown by two parallel lines, one longer than the other. The longer line represents the positive (+) terminal. A true *battery*, which is a combination of two or more cells in series, is indicated by several parallel lines, alternately long and short. It's not necessary to use more than four lines to represent a battery, although you'll often see six, eight, 10, or even 12 lines. Symbols for a cell and a battery are shown in Fig. 1-2.

Meters are portrayed as circles. Sometimes the circle has an arrow inside it, and the meter type, such as mA (milliammeter) or V (voltmeter) is written alongside the circle, as shown in Fig. 1-3A. Sometimes the meter type is indicated inside the circle, and there is no arrow (Fig. 1-3B). It doesn't matter which way you draw them, as long as you're consistent throughout a schematic diagram.

Some other common symbols include the *incandescent lamp*, the *capacitor*, the *air-core coil*, the *iron-core coil*, the *chassis ground*, the *earth ground*, the *AC source*, the set of *terminals*, and the *black box* (general component or device). These are shown in Fig. 1-4.

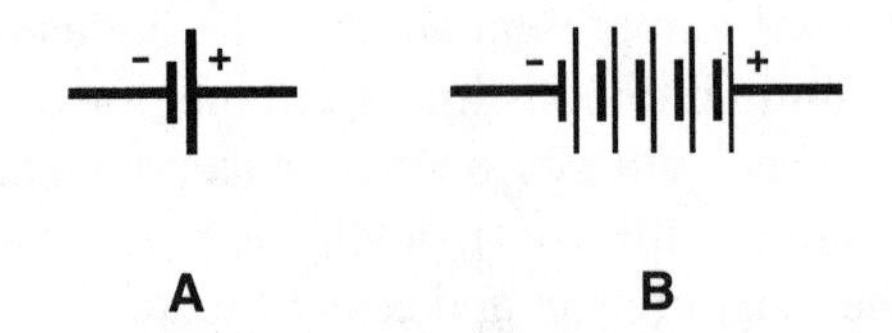

Figure 1-2 Schematic symbols for an electrochemical cell (A) and an electrochemical battery (B).

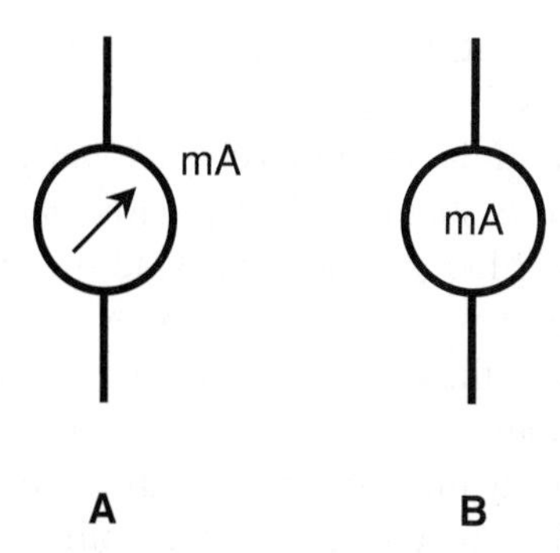

Figure 1-3 Meter symbols can have the designator either outside the circle (A) or inside (B). In this case, both symbols represent a milliammeter (mA).

In a *schematic diagram*, the interconnection of the components is shown, but the actual values of the components are not necessarily indicated. You might see a diagram of a two-transistor audio amplifier, for example, with resistors and capacitors and coils and transistors, but without any data concerning the values or ratings of the components. This is a schematic diagram, but not a *wiring diagram*. It

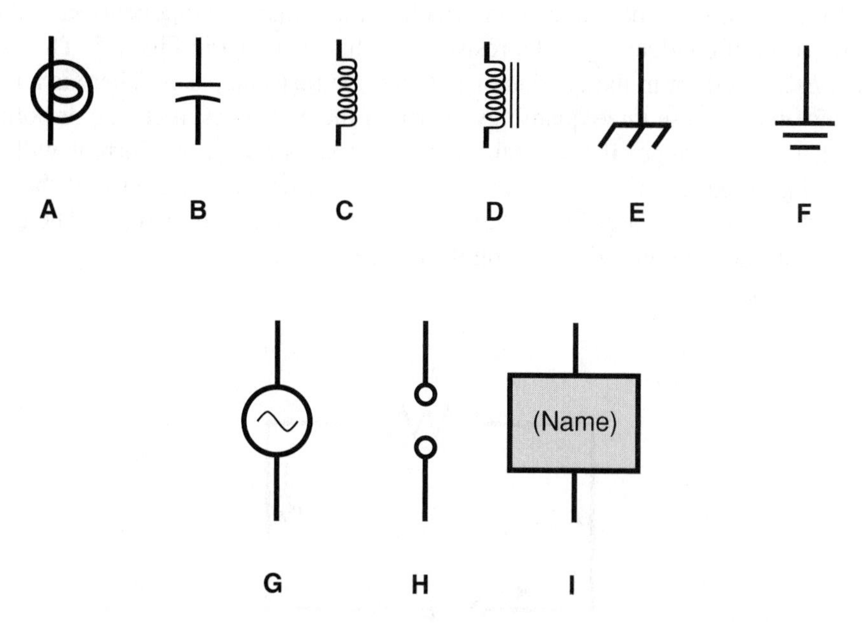

Figure 1-4 Schematic symbols for incandescent lamp (A), fixed capacitor (B), fixed inductor with air core (C), fixed inductor with laminated-iron core (D), chassis ground (E), earth ground (F), signal generator or source of alternating current (G), pair of terminals (H), and specialized component or device (I).

gives the *scheme* for the circuit, but you can't *wire* the circuit and make it work, because there isn't enough information.

Suppose you want to build the circuit. You go to an electronics store to get the parts. What values of resistors should you buy? How about capacitors? What type of transistor will work best? Do you need to wind the coils yourself, or can you get them ready-made? Are there test points or other special terminals that should be installed for the benefit of the technicians who might have to repair the amplifier? How many watts should the potentiometers be able to handle? All these things are indicated in a wiring diagram. You might have seen this kind of diagram in the back of the instruction manual for a hi-fi amplifier, a stereo tuner, or a parametric equalizer. Wiring diagrams are especially useful when you want to build, modify, or repair an electronic device.

Voltage-Current-Resistance Circuits

Most DC circuits can be reduced, in effect, to three major components: a voltage source, a set of conductors, and a resistance. This is shown in Fig. 1-5. The voltage is *E*, the current in the conductor is *I*, and the resistance is *R*. There is a relationship among these three quantities. If one of them changes, then one or both of the others will change. If you make the resistance smaller, the current will get larger. If you reduce the applied voltage, the current will also decrease. If the current in the circuit increases, the voltage across the resistor will increase. There is a simple arithmetic relationship among these three quantities.

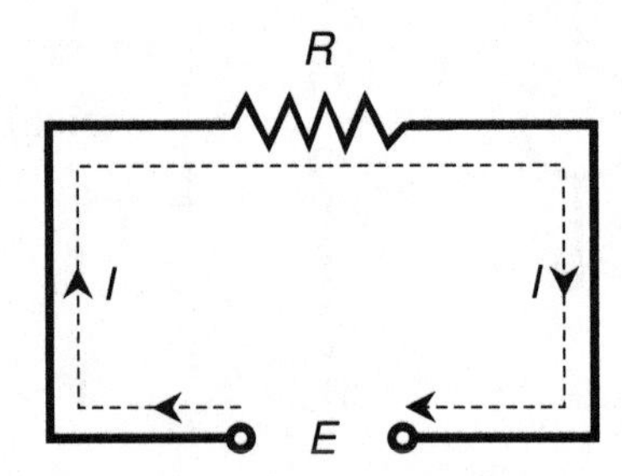

Figure 1-5 The basic elements of a DC circuit. The voltage is *E*, the current is *I*, and the resistance is *R*.

OHM'S LAW

The interdependence among current, voltage, and resistance in DC circuits is called *Ohm's Law*, named after the scientist who first quantified it. Three formulas define it:

$$E = IR$$
$$I = E/R$$
$$R = E/I$$

You need only remember the first formula to derive the others. The easiest way to remember it is to learn the abbreviations E for voltage, I for current, and R for resistance, and then remember that they appear in alphabetical order with the equals sign after the E. Sometimes the three symbols are arranged in the so-called *Ohm's Law triangle*, shown in Fig. 1-6. To find the value of a quantity, cover it up and read the positions of the others.

Remember that you must use the standard units of voltage, current, and resistance for the Ohm's Law formulas to work right. These units are *volts* (V), *amperes* (A), and *ohms* (Ω) respectively. If you use, say, volts and microamperes to calculate a resistance, you cannot be sure of the units you'll end up with when you derive the final result. If the initial quantities are given in units other than volts, amperes, and ohms, convert to these units, and then calculate. After that, you can convert the calculated current, voltage, or resistance value to whatever size unit you want. For example, if you get 13,500,000 Ω as a calculated resistance, you might prefer to say that it's 13.5 *megohms* (MΩ).

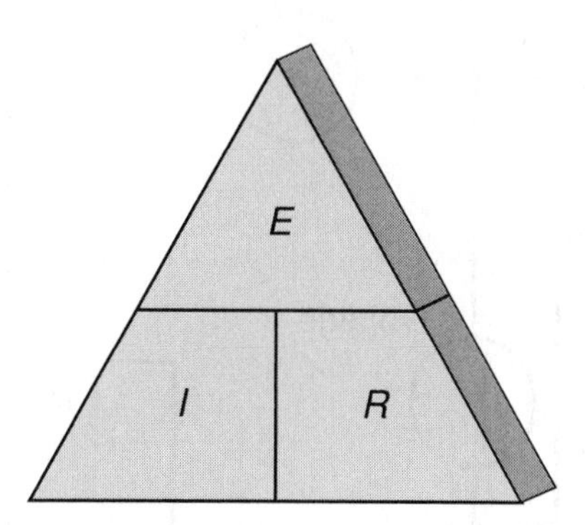

Figure 1-6 The Ohm's Law triangle. The voltage is E, the current is I, and the resistance is R. These quantities are expressed in volts, amperes, and ohms, respectively.

CURRENT CALCULATIONS

Ohm's Law can be used to determine the current in a DC circuit. In order to find the current, you must know the voltage and the resistance, or be able to deduce them. Refer to the schematic diagram of Fig. 1-7. It consists of a DC voltage source, a voltmeter, some wire, an ammeter, and a calibrated, wide-range potentiometer.

PROBLEM 1-1

Suppose that the DC generator in Fig. 1-7 produces 10 V and the potentiometer is set to a value of 10 Ω. What is the current?

SOLUTION 1-1

This is solved by the formula $I = E/R$. Plug in the values for E and R; they are both 10, because the units are given in volts and ohms. Then $I = 10/10 = 1.0$ A.

PROBLEM 1-2

Imagine that the DC generator in Fig. 1-7 produces 100 V and the potentiometer is set to 10 *kilohms* (10 kΩ). What is the current?

SOLUTION 1-2

First, note that 1 kΩ = 1000 Ω. Therefore, you should convert the resistance to ohms and then plug the values in: $I = 100/10{,}000 = 0.01$ A. You can also express this as 10 mA.

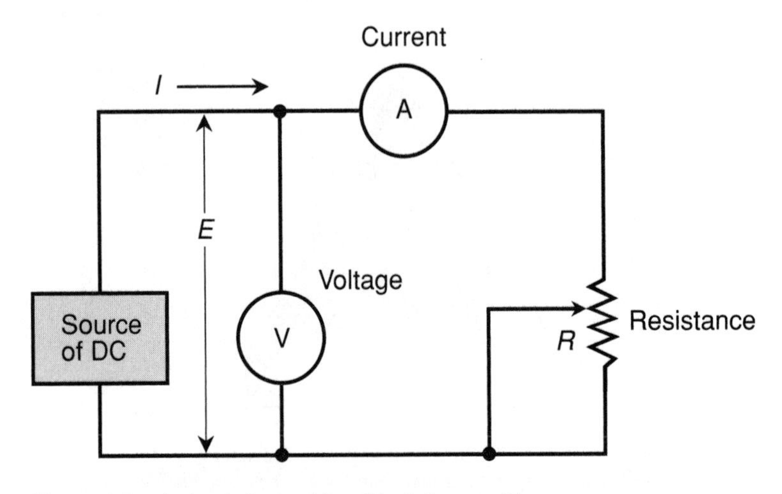

Figure 1-7 A circuit for working Ohm's Law problems.

PROBLEM 1-3
Suppose that the DC generator in Fig. 1-7 is set to provide 88.5 V, and the potentiometer is set to 477 MΩ. What is the current?

SOLUTION 1-3
This problem involves numbers that aren't exactly round, and one of them is huge. But you can use a calculator. First, change the resistance value to ohms, so you get 477,000,000 Ω. Then plug into the Ohm's Law formula: $I = E/R = 88.5 / 477{,}000{,}000 = 0.000000186$ A. It is more reasonable to express this as 0.186 *microamperes* (0.186 μA) or 186 *nanoamperes* (186 nA), where 1 μA = 10^{-6} A and 1 nA = 10^{-9} A.

VOLTAGE CALCULATIONS

Another application of Ohm's Law is to find unknown DC voltages when the current and the resistance are known. Let's work out some problems of this kind.

PROBLEM 1-4
Suppose the potentiometer in Fig. 1-7 is set to 100 Ω, and the measured current is 10 mA. What is the DC voltage?

SOLUTION 1-4
Use the formula $E = IR$. First, convert the current to amperes: 10 mA = 0.01 A. then multiply: $E = 0.01 \times 100 = 1.0$ V.

PROBLEM 1-5
Adjust the potentiometer in Fig. 1-7 to a value of 157 kΩ, and suppose the current reading is 17.0 mA. What is the voltage of the source?

SOLUTION 1-5
You must convert both the resistance and the current values to their proper units. A resistance of 157 kΩ is 157,000 Ω, and a current of 17.0 mA is 0.0170 A. Then $E = IR = 0.017 \times 157{,}000 = 2669$ V = 2.669 kV. You can round this off to 2.67 kV.

PROBLEM 1-6
Suppose you set the potentiometer in Fig. 1-7 so that the meter reads 1.445 A, and you observe that the potentiometer scale shows 99.0 Ω. What is the voltage?

SOLUTION 1-6

These units are both in their proper form. Therefore, you can plug them right in: $E = IR = 1.445 \times 99.0 = 143.055$ V. This can and should be rounded off—but to what extent? This is a good time to state an important rule that should be followed in all technical calculations.

THE RULE OF SIGNIFICANT FIGURES

Competent engineers and scientists go by the *rule of significant figures*, also called the *rule of significant digits*. After completing a calculation, you should always round the answer off to the *least* number of digits given in the input data numbers.

If you follow this rule in Problem and Solution 1-6, you must round off the answer to three significant digits, getting 143 V, because the resistance (99.0 Ω) is only specified to that level of accuracy. If the resistance were given as 99.00 Ω, then you could state the answer as 143.1 V. However, any further precision in the resistance value would not entitle you to go to any more digits in your answer, unless the current were specified to more than four significant figures.

This rule takes some getting-used-to if you haven't known about it or practiced it before. But after awhile, it will become a habit.

RESISTANCE CALCULATIONS

Ohms' Law can be used to find a resistance between two points in a DC circuit, when the voltage and the current are known.

PROBLEM 1-7

If the voltmeter in Fig. 1-7 reads 24 V and the ammeter shows 3.0 A, what is the resistance of the potentiometer?

SOLUTION 1-7

Use the formula $R = E/I$, and plug in the values directly, because they are expressed in volts and amperes: $R = 24/3.0 = 8.0$ Ω. Note that you can specify this value to two significant figures, rather than saying simply 8 Ω. This is because you are given both the voltage and the current to two significant figures. If the ammeter reading had been given as 3 A, you would only be entitled to express the answer as 8 Ω, to one significant digit. The digit 0 is just as important in calculations as any of the other digits 1 through 9.

PROBLEM 1-8

What is the value of the resistance in Fig. 1-7 if the current is 18 mA and the voltage is 229 mV?

SOLUTION 1-8

First, convert these values to amperes and volts. This gives $I = 0.018$ A and $E = 0.229$ V. Then plug into the equation: $R = E/I = 0.229/0.018 = 13\ \Omega$.

PROBLEM 1-9

Suppose the ammeter in Fig. 1-7 reads 52 μA and the voltmeter indicates 2.33 kV. What is the resistance?

SOLUTION 1-9

Convert to amperes and volts, getting $I = 0.000052$ A and $E = 2330$ V. Then plug into the formula: $R = E/I = 2330/0.000052 = 45{,}000{,}000\ \Omega = 45\ \text{M}\Omega$.

POWER CALCULATIONS

You can calculate the power P in a DC circuit such as that shown in Fig. 1-7 by using the formula $P = EI$. This formula tells us that the power in watts (W) is the product of the voltage in volts and the current in amperes. If you are not given the voltage directly, you can calculate it if you know the current and the resistance.

Recall the Ohm's Law formula for obtaining voltage: $E = IR$. If you know I and R but you don't know E, you can get the power P this way:

$$P = EI = (IR)I = I^2R$$

Suppose you're given only the voltage and the resistance. Remember the Ohm's Law formula for obtaining current: $I = E/R$. Therefore:

$$P = EI = E(E/R) = E^2/R$$

PROBLEM 1-10

Suppose that the voltmeter in Fig. 1-7 reads 12 V and the ammeter shows 50 mA. What is the power dissipated by the potentiometer?

SOLUTION 1-10

Use the formula $P = EI$. First, convert the current to amperes, getting $I = 0.050$ A. (Note that the last 0 counts as a significant digit.) Then multiply by 12 V, getting $P = EI = 12 \times 0.050 = 0.60$ W.

PROBLEM 1-11

If the resistance in the circuit of Fig. 1-7 is 999 Ω and the voltage source delivers 3 V, what is the power dissipated by the potentiometer?

SOLUTION 1-11

Use the formula $P = E^2/R = (3 \times 3) / 999 = 9 / 999 = 0.009$ W = 9 mW. You are justified in going to only one significant figure here.

PROBLEM 1-12

Suppose the resistance in Fig. 1-7 is 47 kΩ and the current is 680 mA. What is the power dissipated by the potentiometer?

SOLUTION 1-12

Use the formula $P = I^2R$, after converting to ohms and amperes. Then $P = 0.680 \times 0.680 \times 47{,}000 = 22{,}000$ W = 22 kW. (This is an unrealistic state of affairs: an ordinary potentiometer, such as the type you would use as the volume control in a radio, dissipates 22 kW, several times more than a typical household!)

PROBLEM 1-13

How much voltage would be necessary to drive 521 μA through a resistance of 21.9 kΩ?

SOLUTION 1-13

Use Ohm's Law to find the voltage after converting to amperes and volts: $E = IR = 0.000521 \times 21{,}900 = 11.4$ V.

Resistances in Series and Parallel

When you place resistances in series, their ohmic values add together to get the total resistance. This is easy to remember.

When resistances are placed in parallel, they behave differently than they do in series. One way to look at resistances in parallel is to consider them as *conductances* instead. In parallel, conductances add up directly, just as resistances add in series. If you change all the ohmic values to units of conductance known as *siemens*, you can add these figures up and convert the final answer back to ohms.

The symbol for conductance, when expressed as a mathematical variable, is an uppercase italic *G*. The symbol for the unit, the siemens, is an uppercase non-italic S. Conductance in siemens is related to resistance in ohms by these formulas:

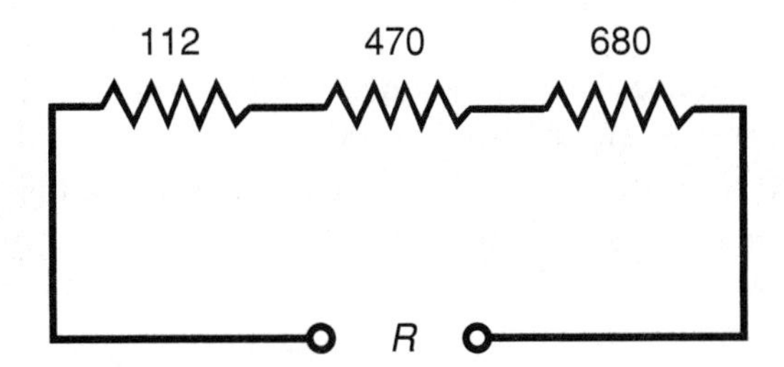

Figure 1-8 Three resistors in series. Illustration for Problem 1-14. Resistance values are in ohms.

$$G = 1/R$$
$$R = 1/G$$

PROBLEM 1-14

Suppose three resistors are connected in series, with the values shown in Fig. 1-8. What is the total resistance of the series combination?

SOLUTION 1-14

Simply add up the values, getting a total of 112 + 470 + 680 = 1262 Ω. You might round this off to 1260 Ω. It depends on the *tolerances* of the resistors—how precise their actual values are to the ones specified by the manufacturer.

PROBLEM 1-15

Consider five resistors in parallel. Call them R_1 through R_5, and call the total resistance R as shown in Fig. 1-9. Let the resistance values be as follows: R_1 = 100 Ω, R_2 = 200 Ω, R_3 = 300 Ω, R_4 = 400 Ω, and R_5 = 500 Ω. What is the total resistance, R, of this parallel combination?

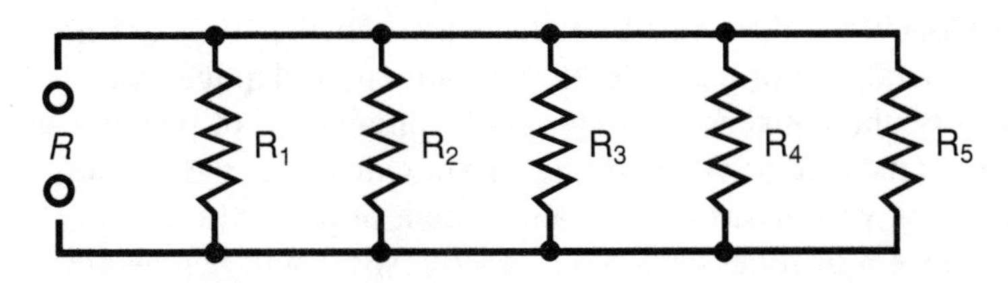

Figure 1-9 Five resistors connected in parallel. Illustration for Problems 1-15 and 1-16.

SOLUTION 1-15

Converting the resistances to conductance values, you get: $G_1 = 1/100 = 0.01$ S, $G_2 = 1/200 = 0.005$ S, $G_3 = 1/300 = 0.00333$ S, $G_4 = 1/400 = 0.0025$ S, and $G_5 = 1/500 = 0.002$ S. Adding these gives $G = 0.01 + 0.005 + 0.00333 + 0.0025 + 0.002 = 0.0228$ S. The total resistance is therefore $R = 1/G = 1/0.0228 = 43.8\ \Omega$.

PROBLEM 1-16

Consider five resistors, called R_1 through R_5, connected in parallel as shown in Fig. 1-9. Suppose all the resistances, R_1 through R_5, are 4.70 kΩ. What is the total resistance, *R*, of this combination?

SOLUTION 1-16

When you have two or more resistors connected in parallel and their resistances are all the same, the total resistance is equal to the resistance of any one component, divided by the number of components. In this example, convert the resistance of any single resistor to 4700 Ω, and then divide this by 5. Thus, you can see that the total resistance is 4700 / 5 = 940 Ω.

In a situation like this, where you have a bunch of resistors connected together to operate as a single unit, the total resistance is sometimes called the *net resistance*. Take note, too, that R is not italicized when it means "resistor," but *R* is italicized when it means "resistance"!

Division of Power

When combinations of resistances are connected to a source of voltage, they draw current. You can figure out how much current they draw by calculating the total resistance of the combination, and then considering the network as a single resistor.

If the resistors in the network all have the same ohmic value, the power from the source is evenly distributed among them, whether they are hooked up in series or in parallel. For example, if there are eight identical resistors in series with a battery, the network consumes a certain amount of power, each resistor bearing 1/8 of the load. If you rearrange the circuit so that the resistors are in parallel, the circuit will dissipate a certain amount of power (a lot more than when the resistors were in series), but again, each resistor will handle 1/8 of the total power load.

Resistances in Series-Parallel

Sets of resistors, all having identical ohmic values, can be connected together in parallel sets of series networks, or in series sets of parallel networks. By doing this, the total power handling capacity of the resistance can be greatly increased over that of a single resistor.

Sometimes, the total resistance of a *series-parallel network* is the same as the value of any one of the resistors. This is always true if the components are identical, and are in a network called an *n-by-n matrix*. That means, when n is a whole number, there are n parallel sets of n resistors in series (Fig. 1-10A), or else there are n series sets of n resistors in parallel (Fig. 1-10B). Either arrangement gives the same practical result.

Engineers sometimes use series-parallel networks to obtain resistances with large power-handling capacity. A series-parallel array of n by n resistors will have n^2 times the power-handling capacity, in watts (W), of a single resistor. For example, a 3×3 series-parallel matrix of 2-W resistors can handle up to $3^2 \times 2 = 9 \times 2 = 18$ W. A 10×10 array of 1-W resistors can dissipate up to $10^2 \times 1 = 100$ W. The total power-handling capacity is multiplied by the total number of resistors in

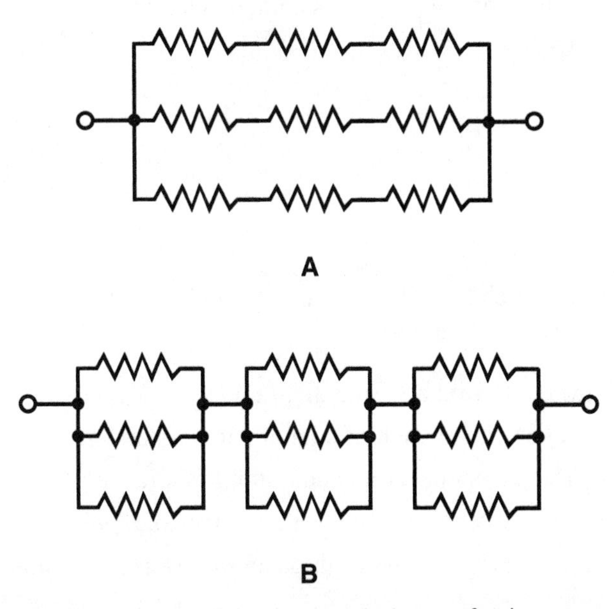

Figure 1-10 Series-parallel resistances. At A, sets of series-connected resistors are in parallel. At B, sets of parallel-connected resistances are in series. These examples show symmetrical n-by-n matrices with $n = 3$.

the matrix. But this is true only if all the resistors have the same ohmic values, and the same power-dissipation ratings.

It is unwise to build series-parallel arrays from resistors with different ohmic values or power ratings. If the resistors have values and/or ratings that are the slightest bit nonuniform, one of them might be subjected to more current than it can withstand, and then it will burn out. After that happens, the current distribution in the network can change so a second component fails, and then a third. It's difficult to predict the current and power distribution in an array when its resistor values are all different.

If you need a resistance with a certain power-handling capacity, you must be sure the network can handle at least that much power. If a 50-W rating is required, and a certain combination will handle 75 W, that's fine. But it isn't good enough to build a circuit that will handle only 48 W. Some extra tolerance, say 10 percent over the minimum rating needed, is good, but it's silly to make a 500-W network, using far more resistors than necessary.

Non-symmetrical series-parallel networks, made up from identical resistors, can increase the power-handling capability over that of a single resistor. But in these cases, the total resistance is not the same as the value of the single resistors. The overall power-handling capacity is always multiplied by the total number of resistors, whether the network is symmetrical or not, provided all the ohmic values are identical. In engineering work, cases sometimes arise where non-symmetrical networks fit the need.

Quiz

Refer to the text in this chapter if necessary. A good score is at least 8 correct. The answers are in the back of the book.

1. You can expect to find a wiring diagram
   (a) on a sticker on the back of a parametric equalizer.
   (b) in an advertisement for an audio amplifier.
   (c) in the service/repair manual for a microphone preamplifier.
   (d) in the photograph of the front panel of a stereo hi-fi tuner.

For questions 2 through 5, please refer to Fig. 1-7. Remember to take significant figures into account when completing your calculations!

2. Suppose the resistance is 472 Ω, and the current is 875 mA. The source voltage must therefore be
   (a) 413 V.
   (b) 0.539 V.
   (c) 1.85 V.
   (d) None of the above

3. Given a DC voltage source of 3.5 kV and a resistance of 220 Ω, what is the current?
   (a) 16 mA.
   (b) 6.3 mA.
   (c) 6.3 A.
   (d) None of the above

4. A source delivers 12 V and the current is 777 mA. The best expression for the resistance is
   (a) 15 Ω.
   (b) 15.4 Ω.
   (c) 9.3 Ω.
   (d) 9.32 Ω.

5. Suppose the voltage from the source is 12 V and the potentiometer is set for 470 Ω. The power dissipated in the resistance is approximately
   (a) 0.31 W.
   (b) 25.5 mW.
   (c) 39.2 W.
   (d) 3.26 W.

6. Suppose six resistors are hooked up in series, and each of them has a value of 540 Ω. What is the resistance across the entire combination?
   (a) 90 Ω.
   (b) 3.24 kΩ.
   (c) 540 Ω.
   (d) None of the above

7. Suppose you have three resistors in parallel, each with a value of 69 kΩ. The net resistance of the combination is
   (a) 23 Ω.
   (b) 23 kΩ.
   (c) 204 Ω.
   (d) 0.2 MΩ.

8. Suppose you have an unlimited supply of 100-Ω, 1-W resistors. You need a 100-Ω resistance capable of dissipating at least 10 W. This can be done most cheaply, using only these 100 Ω, 1-W resistors, by constructing a
   (a) 3 × 3 series-parallel matrix.
   (b) 4 × 3 series-parallel matrix.
   (c) 4 × 4 series-parallel matrix.
   (d) 2 × 5 series-parallel matrix.

9. Suppose you have an unlimited supply of 1000-Ω, 1-W resistors, and you need a 500-Ω resistance rated at a minimum of 7 W. This can be done by assembling
   (a) four sets of two resistors in series, and connecting these four sets in parallel.
   (b) four sets of two resistors in parallel, and connecting these four sets in series.
   (c) a 3 × 3 series-parallel matrix of resistors.
   (d) a series-parallel matrix, but something different than those described above.

10. Good engineering requires that a series-parallel resistive network be assembled
    (a) from resistors that are all different.
    (b) from resistors that are all identical.
    (c) from a series combination of resistors in parallel but not from a parallel combination of resistors in series.
    (d) from a parallel combination of resistors in series, but not from a series combination of resistors in parallel.

CHAPTER 2

Alternating Current Basics

Direct current can be expressed in terms of two variables: direction (polarity) and intensity (amplitude). Alternating current (AC) is a little more complicated. This chapter will acquaint you with some common forms of AC.

Period and Frequency

In AC, the *polarity* reverses at regular intervals. The *instantaneous amplitude* (the current or voltage at any given *instant* in time) of AC usually varies because of the repeated reversal of polarity. But there are certain cases where the amplitude remains constant, even though the polarity keeps reversing. The rate of change of polarity is the variable that makes AC so much different from DC. The behavior of an AC wave depends largely on this rate: the *frequency*.

In a *periodic AC wave*, the function of *instantaneous amplitude vs. time* repeats indefinitely. The length of time between one repetition of the pattern, or one *cycle*, and the next is called the *period* of the wave. This is illustrated in Fig. 2-1 for a simple AC wave. The period is the time between points at identical positions on two successive wave cycles. When expressed in common time units such as seconds (s), milliseconds (ms), or microseconds (μs), the period is denoted by *T*.

In the olden days, AC frequency was specified in *cycles per second*, abbreviated *cps*. High frequencies were expressed in units of *kilocycles*, *megacycles*, or *gigacycles*, representing thousands, millions, or billions (thousand-millions) of cycles per second. Nowadays, the cycle per second is known as the *hertz*, abbreviated *Hz*. Thus, 1 Hz = 1 cps, 10 Hz = 10 cps, and so on. Higher frequencies are given in *kilohertz* (kHz), *megahertz* (MHz), or *gigahertz* (GHz). The relationships are:

$$1 \text{ kHz} = 1000 \text{ Hz}$$
$$1 \text{ MHz} = 1000 \text{ kHz} = 1{,}000{,}000 \text{ Hz} = 10^6 \text{ Hz}$$
$$1 \text{ GHz} = 1000 \text{ MHz} = 1{,}000{,}000{,}000 \text{ Hz} = 10^9 \text{ Hz}$$

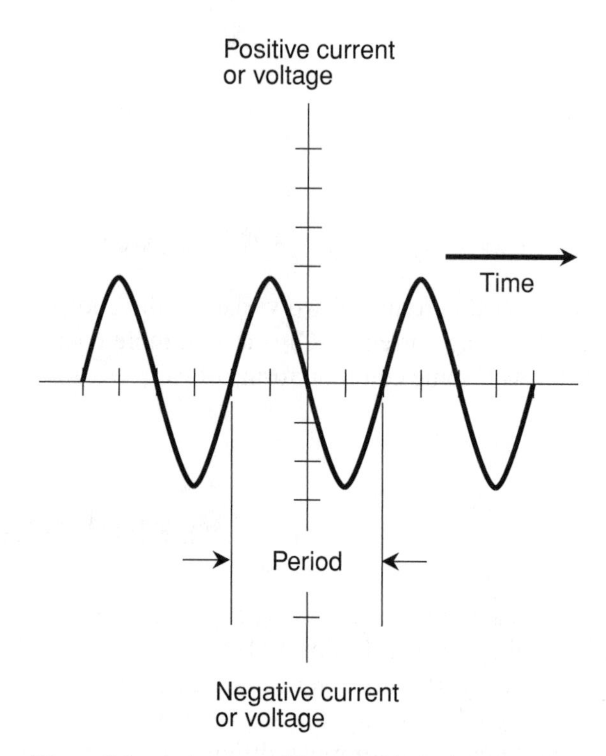

Figure 2-1 A sine wave. The period is the length of time it takes for one cycle to be completed.

Sometimes an even bigger unit, the *terahertz* (THz), is used to specify AC frequency. This is a trillion (1,000,000,000,000 or 10^{12}) hertz. Electrical currents generally do not attain such frequencies, although some forms of *electromagnetic radiation* do. *Audio-frequency* (AF) currents have frequencies ranging from about 20 Hz to 20 kHz, the same as the range of acoustic frequencies that humans can hear.

The frequency of an AC wave, denoted *f*, in hertz is the reciprocal of the period in seconds. Mathematically, these two equations express the relationship:

$$f = 1/T$$

and

$$T = 1/f$$

Some AC waves have only one frequency component. These waves are called *pure*. But often, there are components at multiples of the main, or *fundamental*, frequency. There can also be components at "odd" frequencies. Some AC waves have hundreds, thousands, or even infinitely many different component frequencies.

PROBLEM 2-1

What is the frequency, in hertz, of a wave that has a period of 5 ms? Note that 1 ms = 0.001 s.

SOLUTION 2-1

Use the formula for frequency in terms of period, making sure to express the period in seconds, as follows:

$$\begin{aligned} f &= 1/T \\ &= 1/0.005 \\ &= 200 \text{ Hz} \end{aligned}$$

PROBLEM 2-2

What is the period, in milliseconds, of a wave that has a frequency of 50 Hz?

SOLUTION 2-2

In this case, you should use the formula for period in terms of frequency:

$$\begin{aligned} T &= 1/f \\ &= 1/50 \\ &= 0.02 \text{ s} \\ &= 200 \text{ ms} \end{aligned}$$

Waveforms

The amplitude-vs.-time function of an AC wave can have an infinite variety of different shapes. The shape of a periodic wave (one that repeats at regular intervals) is known as its *waveform*. Here are some of the most common waveforms encountered in audio electronics.

THE SINE WAVE

Sometimes, alternating current has a *sine-wave*, or *sinusoidal*, nature. This means that the direction of the current reverses at regular intervals, and that the voltage- or current-vs.-time curve is shaped like the trigonometric *sine function*. The waveform illustrated in Fig. 2-1 is a sine wave. Any AC wave that concentrates all its energy at a single frequency has a perfectly sinusoidal shape. Conversely, any perfect sinusoidal AC wave has a single defined frequency.

SQUARE WAVES

Strangely enough, there can be an AC wave whose instantaneous amplitude remains constant, even though the polarity reverses. A *square wave* is such a wave.

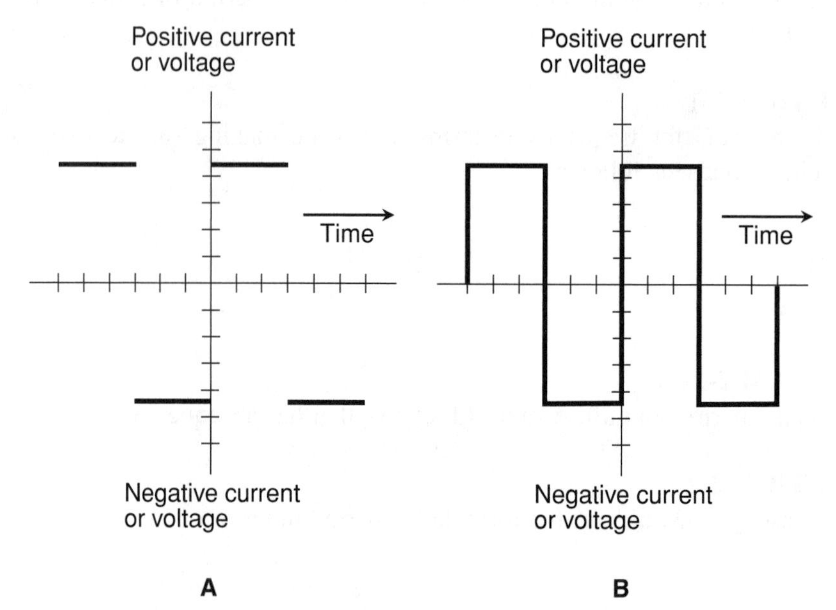

Figure 2-2 At A, a perfect square wave; the transitions are instantaneous and therefore do not show up on the graph. At B, the more common rendition of a square wave, showing the transitions as vertical lines.

When displayed on an *oscilloscope*, a square wave looks like a pair of parallel, dashed lines, one with positive polarity and the other with negative polarity (Fig. 2-2A). The oscilloscope display is in effect a *rectangular-coordinate graph* that shows voltage on the vertical scale and time on the horizontal scale. The transitions between negative and positive for a theoretically perfect square wave would not show up on the oscilloscope, because they would be instantaneous. But in practice, the transitions can often be seen as vertical lines (Fig. 2-2B).

True square waves have equal negative and positive peaks. Thus, the absolute amplitude of the wave is constant. Half of the time it's $+x$, and the other half of the time it's $-x$ (where x can be expressed in volts, amperes, or watts). Some squared-off waves are lopsided; the negative and positive amplitudes are not the same. Still others remain at positive polarity longer than they remain at negative polarity (or vice-versa). These are examples of *asymmetrical square waves*, also called *rectangular waves.*

SAWTOOTH WAVES

Some AC waves rise and/or fall in straight, sloping lines as seen on an oscilloscope screen. The slope of the line indicates how fast the voltage or current is changing. Such waves are called *sawtooth waves* because of their appearance. Sawtooth waves are generated by certain electronic test devices. They can also be generated by electronic sound synthesizers.

- Fast-rise, slow-decay: Figure 2-3 shows a sawtooth wave in which the positive-going slope (called the *rise*) is extremely steep, as with a square wave, but the negative-going slope (called the *decay*) is not so steep.

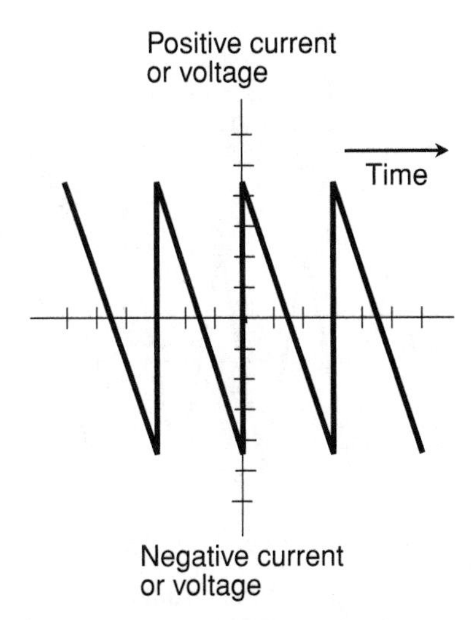

Figure 2-3 A sawtooth wave with a near-zero rise time and a finite decay time.

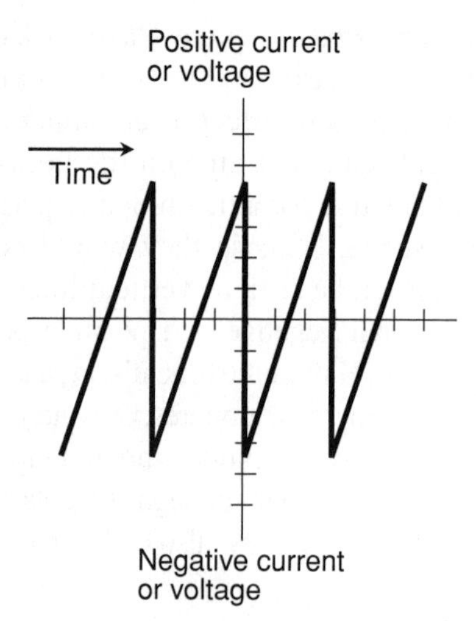

Figure 2-4 A sawtooth wave with a finite rise time and a near-zero decay time.

- Slow-rise, fast-decay: Another form of sawtooth wave is just the opposite, with a defined, nonzero rise time and a near-zero decay time. This type of wave is also called a *ramp* because the sloped part of it looks like an incline going upwards (Fig. 2-4).
- Variable rise and decay: Sawtooth waves can have rise and decay slopes in an infinite number of different combinations. One common example is shown in Fig. 2-5. In this case, the rise and the decay times are both nonzero and equal. This is known as a *triangular wave*.

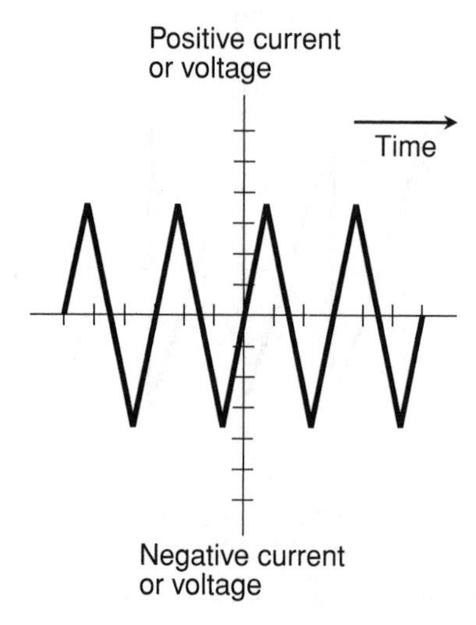

Figure 2-5 A triangular wave with equal, finite rise and decay times.

COMPLEX AC WAVES

As long as an electrical wave has a definable period, and as long as the polarity keeps switching back and forth between positive and negative, it is AC, no matter how complicated or bizarre the waveform happens to be. Figure 2-6 shows an example of a *complex AC wave*. There is a definable period, and therefore a definable frequency. The period is the time between identical points on any two successive cycles.

With some waves, it can be difficult or almost impossible to ascertain the period. This is because the wave has two or more components that are of nearly the same amplitude. When this happens, the *frequency spectrum* of the wave is multifaceted. That means the wave energy is split up more or less equally among multiple frequencies.

If an electrical current varies in a manner so complicated that no period exists or can be defined, then that wave is technically a form of *electrical noise*. If electrical noise is input to an amplifier and fed to a speaker and the result is audible, then the sound energy thus produced is a form of *acoustic noise*.

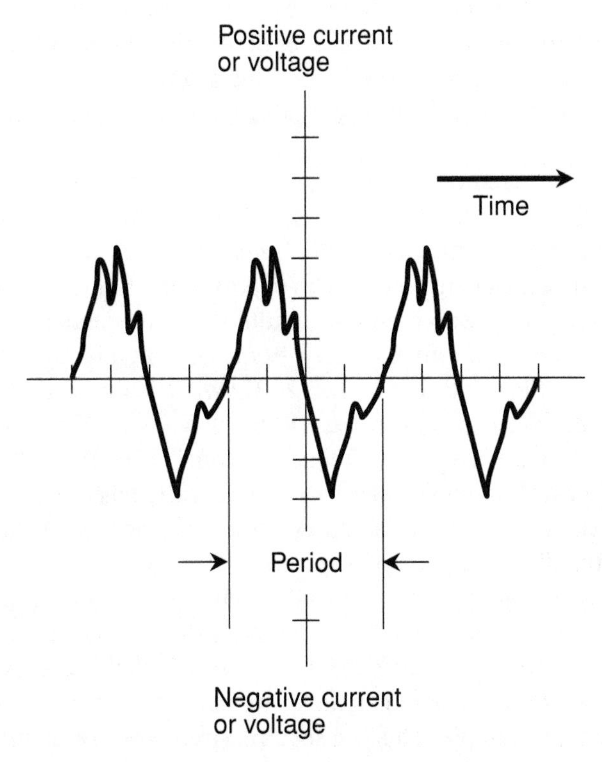

Figure 2-6 An irregular waveform.

PROBLEM 2-3

What happens when two waveforms that have the same frequency, but much different shapes, are combined?

SOLUTION 2-3

When two AC waves are combined, the instantaneous amplitude of the *composite wave* at each point in time is equal to the sum of the instantaneous amplitudes of the two waves at that point in time. The exact shape of the composite wave depends on the shapes of the *component waves*, how strong they are, and whether or not their cycles are "in sync."

FREQUENCY SPECTRUM

On an oscilloscope display, time is rendered on the horizontal axis and represents the *independent variable* or *domain* of the wave function. For this reason, an oscilloscope is said to be a *time-domain* instrument. But suppose you want to see the amplitude of a complex signal as a function of frequency, rather than as a function of time? This can be done with a lab instrument called a *spectrum analyzer*, which provides a *frequency-domain* display. Its horizontal axis shows frequency as the independent variable, ranging from some adjustable minimum frequency (at the extreme left) to some adjustable maximum frequency (at the extreme right).

An AC sine wave, as displayed on a spectrum analyzer, appears as a single *pip*, or vertical line (Fig. 2-7A). This means that all of the energy in the wave is concentrated at one frequency, called the *fundamental frequency*. But most AC waves contain *harmonic* energy along with energy at the fundamental frequency. A harmonic frequency is a whole-number multiple of the fundamental frequency. For example, if 60 Hz is the fundamental frequency, then harmonics can exist at 120 Hz, 180 Hz, 240 Hz, and so on. The 120 Hz wave is the *second harmonic*, the 180 Hz wave is the *third harmonic*, the 240 Hz wave is the *fourth harmonic*, and so on. In general, if a wave has a frequency equal to n times the fundamental (where n is some whole number), then that wave is called the *nth harmonic*. In Fig. 2-7B, a 60 Hz AC wave is shown along with three harmonics, as the combination would look on the display screen of a spectrum analyzer.

Any wave that is not a perfect sine wave contains harmonic energy in addition to energy at the fundamental frequency. The exact shape of a wave depends on the amount of energy in the harmonics, and the way in which this energy is distributed among them. *Irregular AC waves* can have any imaginable frequency distribution. Figure 2-8 shows an example. This is a spectral (frequency-domain) display of an

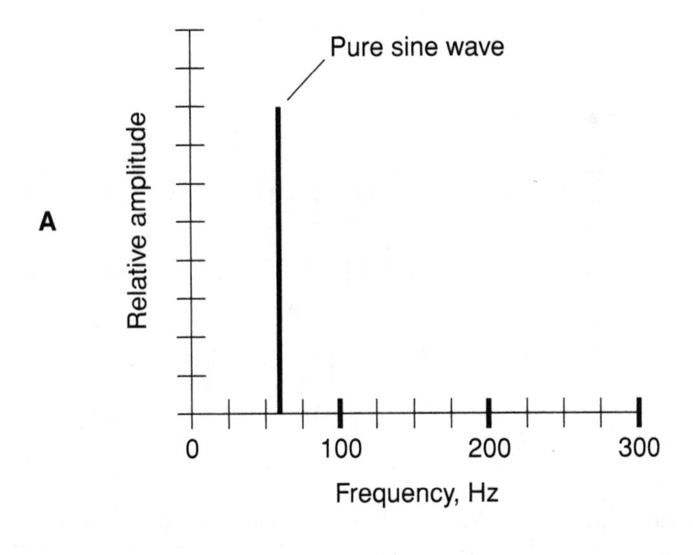

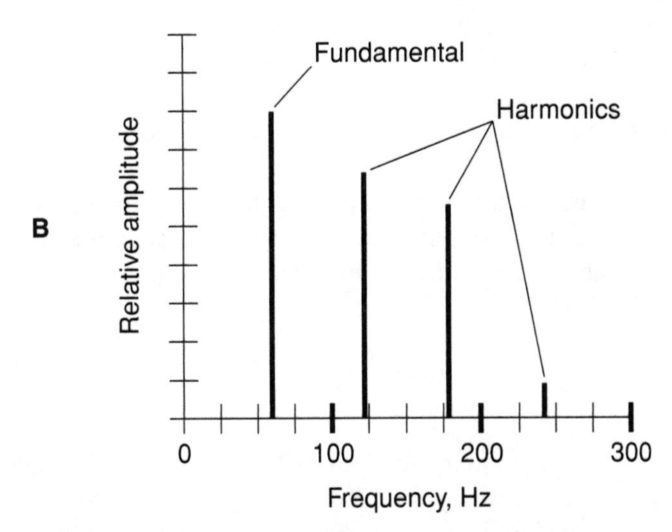

Figure 2-7 At A, a spectral diagram of a pure, 60-Hz sine wave. At B, a spectral diagram of a 60-Hz wave with three harmonics.

amplitude-modulated (AM) voice radio signal. In this example, each horizontal division represents a frequency increment of 1000 Hz. Much of the energy is concentrated at the center of the pattern, at the frequency shown by the vertical line. That is the *carrier frequency*. There is also some signal energy near, but not exactly at, the carrier frequency. That's the part of the signal that contains the voice, and it consists of energy at audio frequencies between approximately 300 Hz and 3000 Hz.

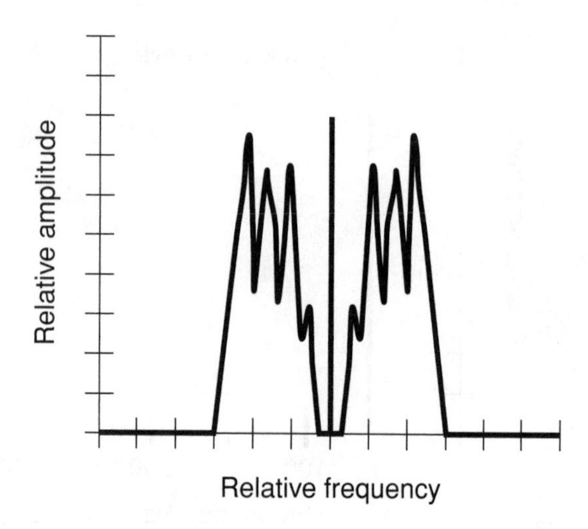

Figure 2-8 A spectral diagram of an amplitude-modulated radio signal. The modulating signal is voice audio, with the frequency range restricted for communications. Each horizontal division represents 1000 Hz.

Fractions of a Cycle

The AC cycle can be broken down into small parts for analysis. One complete cycle can be compared to a single revolution around a circle. Fractions of a cycle are like angles that can be expressed in *degrees* or *radians*, just like angles in geometry and trigonometry.

DEGREES

One method of specifying the phase of an AC cycle is to divide it into 360 equal parts, called *degrees of phase*, symbolized by a superscript, lowercase letter o (°). The value 0° is assigned to the point in the cycle where the amplitude is zero and positive-going. The same point on the next cycle is given the value 360°. The point 1/4 of the way through the cycle is 90°; the point halfway through the cycle is 180°; the point 3/4 of the way through the cycle is 270°. This is illustrated in Fig. 2-9. Degrees of phase are used by engineers and technicians.

RADIANS

The other method of specifying phase is to divide the cycle into 2π equal parts, where π (pi) is a geometric constant equal to the number of diameters of a circle

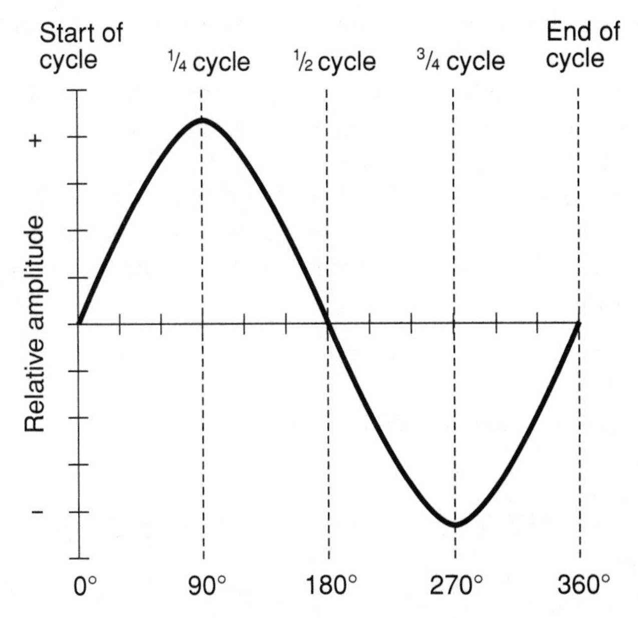

Figure 2-9 A cycle is divided into 360 equal parts, called degrees.

that can be laid end-to-end around the circumference of that circle. This constant is approximately equal to 3.1416. A *radian of phase* (rad) is equal to about 57.3°.

Sometimes, the frequency of an AC wave is measured in *radians per second* (rad/s) rather than in hertz. Because there are 2π (about 6.2832) radians in a complete cycle of 360°, the *angular frequency* of a wave, in radians per second, is equal to approximately 6.2832 times the frequency in hertz. Radians of phase are used by theoretical physicists.

PHASE DIFFERENCE

Even if two AC waves have exactly the same frequency, they can produce different effects because they are "out of sync" with each other. This is especially true when AC waves are added together to produce a third, or *composite*, wave.

If two pure AC sine waves have identical frequencies and identical amplitudes but differ in phase by 180° (a half cycle), they cancel each other out, and the composite wave is zero; it ceases to exist! If the two waves are exactly in phase, the composite wave has the same frequency, but twice the amplitude, of either signal alone.

If two pure AC sine waves have the same frequency but different amplitudes, and if they differ in phase by 180°, the composite signal has the same frequency as the originals, and an amplitude equal to the difference between the two. If two such waves are exactly in phase, the composite has the same frequency as the originals, and an amplitude equal to the sum of the two.

If two pure AC sine waves have the same frequency but differ in phase by some odd amount such as 75° or 110°, the resulting signal has the same frequency, but does not have the same waveform as either of the original signals. The variety of such cases is infinite.

PROBLEM 2-4

How many degrees of phase are represented by 1/6 of a cycle?

SOLUTION 2-4

A complete cycle is 360°. Therefore, 1/6 of a cycle is equal to 360° divided by 6, or 60° of phase.

PROBLEM 2-5

What fraction of a wave cycle is represented by $\pi/8$ radians of phase?

SOLUTION 2-5

A complete cycle is 2π rad. A phase angle of $\pi/8$ rad is equal to $2\pi/16$ rad, so it is 1/16 of a cycle.

PROBLEM 2-6

What is the angular frequency, in radians per second, of a wave with a period of 80 ms? Round the answer off to two significant digits.

SOLUTION 2-6

First, convert the period into frequency in the conventional units (hertz). Note that 80 ms is equal to 0.080 s. Use the formula for frequency in terms of period:

$$\begin{aligned} f &= 1/T \\ &= 1/0.080 \\ &= 12.5 \text{ Hz} \end{aligned}$$

In order to get the angular frequency in radians per second, we must multiply the frequency in hertz by 2π, or approximately 6.2832. Therefore, the angular frequency is about 12.5×6.2832, or 78.54 rad/s. This rounds off to 79 rad/s.

Expressions of Amplitude

Amplitude is also called *magnitude*, *level*, *strength*, or *intensity*. The amplitude of an AC wave is usually specified in microvolts, millivolts, or volts (for voltage) or in microamperes, milliamperes, or amperes (for current). Occasionally it is specified in microwatts, milliwatts, or watts (for power). There are several semantical ways in which amplitude can be defined.

INSTANTANEOUS

The *instantaneous amplitude* of an AC wave is the amplitude at some precise moment, or instant, in time. This value constantly changes. The manner in which it varies depends on the waveform. Instantaneous amplitudes are represented in time-domain graphs or displays by single points on wave curves.

PEAK

The *peak* (pk) *amplitude* of an AC wave is the maximum extent, either positive or negative, that the instantaneous amplitude attains. In many situations, the positive and negative peak amplitudes of an AC wave are the same. But sometimes they differ. Figure 2-9 is an example of a wave in which the positive peak amplitude is the same as the negative peak amplitude. Figure 2-10 is an illustration of a wave that has different positive and negative peak amplitudes.

PEAK-TO-PEAK

The *peak-to-peak* (pk-pk) *amplitude* of a wave is the net difference between the positive peak amplitude and the negative peak amplitude (Fig. 2-11). The peak-to-peak amplitude is equal to the positive peak amplitude plus the negative peak amplitude. When the positive and negative peak amplitudes of an AC wave are equal, the peak-to-peak amplitude is exactly twice the peak amplitude. In the expression of peak-to-peak values, polarity is irrelevant.

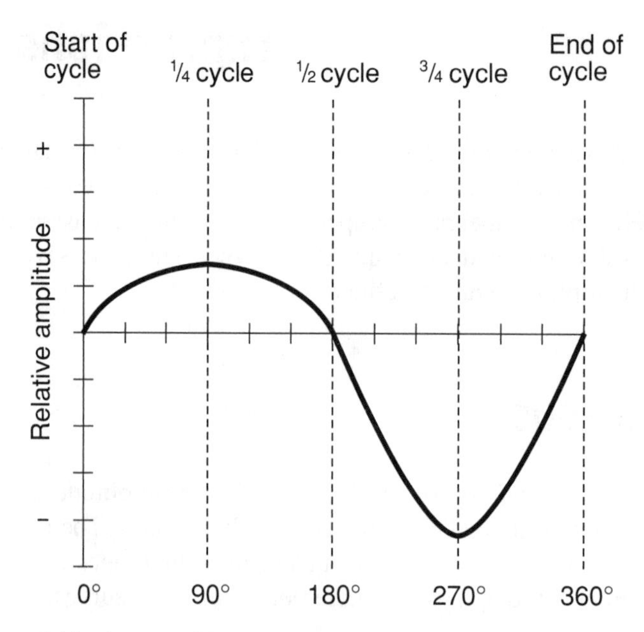

Figure 2-10 A wave with unequal positive and negative peak amplitudes.

ROOT-MEAN-SQUARE (RMS)

Often, it is necessary to express the *effective amplitude* of an AC wave. This is the voltage, current, or power that a DC source would have to produce in order to have the same general effect as a given AC wave. The most common expression for effective AC intensity is called the *root-mean-square*, or rms, amplitude. The terminology reflects the fact that the AC wave is mathematically "operated on" by taking the square root of the mean (average) of the square of all its instantaneous amplitudes.

For a perfect AC sine wave with no DC component, the rms current or voltage is equal to 0.707 times the peak current or voltage. That's 0.354 times the peak-to-peak current or voltage. Conversely, the peak current or voltage is 1.414 times the rms current or voltage.

For a perfect square wave, the rms value is half the peak-to-peak value. For sawtooth and irregular waves, the relationship between the rms value and the peak value depends on the exact shape of the wave. But the rms value is never greater than the peak value for any type of AC wave, as long as there is no DC component. In the expression of rms values, as with peak-to-peak values, polarity is irrelevant.

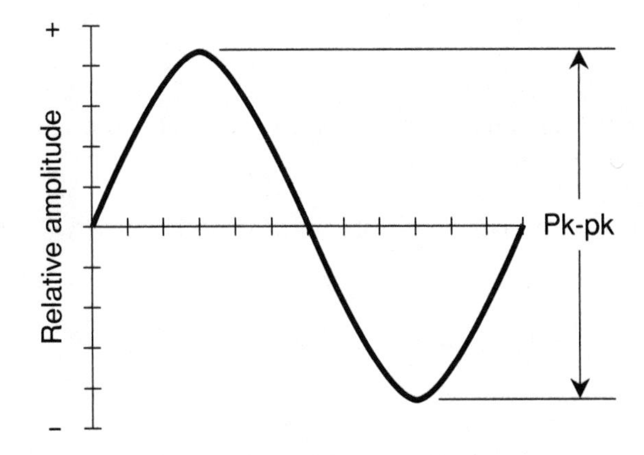

Figure 2-11 Peak-to-peak (pk-pk) amplitude of a sine wave.

PROBLEM 2-7

Suppose a perfect sine wave has a positive peak voltage of +6.5 V and a negative peak voltage of –3.5 V. What is the peak-to-peak voltage?

SOLUTION 2-7

The peak-to-peak voltage is the net difference between the positive peak voltage and the negative peak voltage. Call the positive peak voltage $V_{pk+} = +6.5$ V, and call the negative peak voltage $V_{pk-} = -3.5$ V. Then the peak-to-peak voltage, V_{pk-pk}, is found by subtracting the second of these from the first:

$$\begin{aligned} V_{pk-pk} &= V_{pk+} - V_{pk-} \\ &= +6.5 - (-3.5) \\ &= 6.5 + 3.5 \\ &= 10 \text{ V pk-pk} \end{aligned}$$

PROBLEM 2-8

What is the rms voltage of the sine wave described in Problem 2-7?

SOLUTION 2-8

Let's call the rms voltage V_{rms}. Then:

$$\begin{aligned} V_{rms} &= 0.354\ V_{pk-pk} \\ &= 0.354 \times 10 \\ &= 3.54 \text{ V rms} \end{aligned}$$

Quiz

Refer to the text in this chapter if necessary. A good score is at least 8 correct. Answers are in the back of the book.

1. Which of the following can vary with AC, but never with DC?
   (a) Power.
   (b) Voltage.
   (c) Frequency.
   (d) Amplitude.
2. The length of time between a point in one cycle and the same point in the next cycle of an AC wave is the
   (a) frequency.
   (b) magnitude.
   (c) period.
   (d) polarity.
3. On the display of a spectrum analyzer, an AC signal having only one frequency component looks like
   (a) a single pip.
   (b) a sine wave.
   (c) a square wave.
   (d) sawtooth wave.
4. The period of an AC wave, in seconds, is
   (a) the same as the frequency in hertz.
   (b) not related to the frequency in any way.
   (c) equal to 1 divided by the frequency in hertz.
   (d) equal to the peak amplitude in volts divided by the frequency in hertz.
5. The sixth harmonic of an AC wave whose period is 1.000 millisecond (1.000 ms) has a frequency of
   (a) 0.006 Hz.
   (b) 167 Hz.
   (c) 7 kHz.
   (d) 6 kHz.

6. A degree of phase represents approximately
   (a) 6.28 cycles.
   (b) 57.3 cycles.
   (c) 0.0167 cycle.
   (d) 0.00278 cycle.

7. Suppose that two AC waves have the same frequency but differ in phase by exactly 1/20 of a cycle. The phase difference between these two waves is
   (a) 18°.
   (b) 20°.
   (c) 36°.
   (d) 5.73°.

8. Suppose an AC signal has a frequency of 1770 Hz. The angular frequency is about
   (a) 1770 rad/s.
   (b) 11,120 rad/s.
   (c) 282 rad/s.
   (d) impossible to determine from the data given.

9. A triangular wave exhibits
   (a) zero rise time and nonzero decay time.
   (b) nonzero rise time and zero decay time.
   (c) nonzero rise time and a nonzero decay time.
   (d) zero rise time and zero decay time.

10. If two perfect sine waves have the same frequency and the same amplitude, but are in opposite phase, the composite wave
    (a) has twice the amplitude of either input wave alone.
    (b) has half the amplitude of either input wave alone.
    (c) is complex, but has the same frequency as the originals.
    (d) has zero amplitude.

Fundamentals of Phase

In an AC wave or signal, each 360° cycle is exactly the same as every other. That is, in every cycle, the waveform of the previous cycle is repeated. In this chapter, you'll learn about the ideal AC waveform: the *sine wave*.

The Nature of a Cycle

An AC sine wave has a characteristic shape, as shown in Fig. 3-1. This is the way the graph of the sine function, $y = \sin x$, looks on an (x,y) coordinate plane. (The abbreviation *sin* stands for *sine* in trigonometry.) Suppose the peak voltage of a sine wave is plus or minus one volt (±1 V), as shown, and the period is 1 s, so the frequency is 1 Hz. Consider the wave cycle to begin at time $t = 0$. Then in this example, a cycle begins every time the value of t is a whole number. At every such instant, the voltage is zero and *positive-going*.

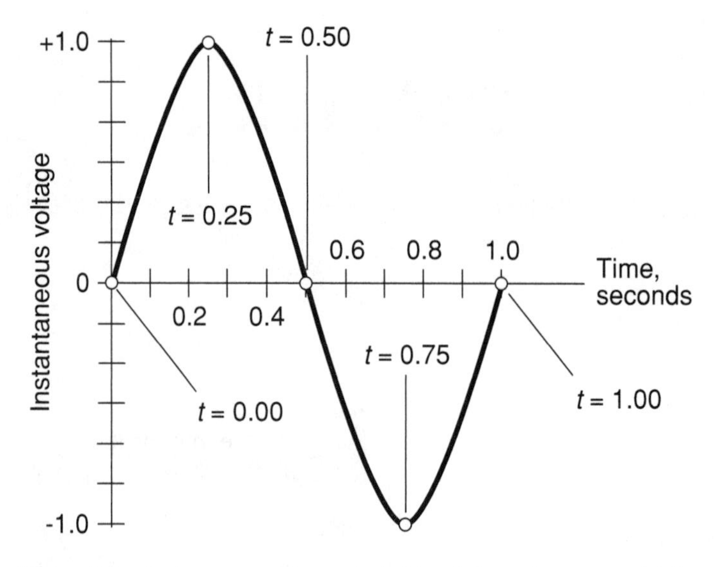

Figure 3-1 A sine wave with a period of 1 s. It has a frequency of 1 Hz.

POINTS IN TIME

If you freeze time at, say, $t = 446.00$, the voltage is zero. Looking at the diagram, you can see that the voltage is also zero every so-many-and-a-*half* seconds, so it is zero at $t = 446.5$. But instead of getting more positive at these middle-of-the-second instants, the voltage is *negative-going*.

If you freeze time at so-many-and-a-*quarter* seconds, say $t = 446.25$, the voltage is +1 V. The wave is exactly at its positive peak. If you stop time at so-many-and-*three-quarter* seconds, say $t = 446.75$, the voltage is exactly at its negative peak, −1 V. At intermediate times, say, so-many-and-*three-tenths* seconds, the voltage has intermediate values.

INSTANTANEOUS RATE OF CHANGE

Figure 3-1 shows that there are instants in time at which the voltage is increasing, and instants in time at which the voltage is decreasing. *Increasing*, in this context, means "getting more positive," and *decreasing* means "getting more negative." The most rapid increase in voltage occurs when $t = 0.0$ and $t = 1.0$. The most rapid

decrease takes place when $t = 0.5$. When $t = 0.25$, and also when $t = 0.75$, the instantaneous voltage neither increases nor decreases. The single, complete cycle shown in Fig. 3-1 represents every possible condition of the AC sine wave having a frequency of 1 Hz and peak voltages of ±1 V. The whole wave recurs, over and over, for as long as the AC continues to flow in the circuit.

The *instantaneous rate of change* of a quantity is defined as the rate at which the value of the quantity changes at some point (instant) in time. Now imagine that you want to observe the instantaneous rate of change in the voltage of the wave in Fig. 3-1, as a function of time. A graph of this turns out to be a sine wave having the same frequency as the original, but it displaced to the left by 1/4 of a cycle. If you plot the instantaneous rate of change of a sine wave against time (Fig. 3-2), you get the *derivative* of the waveform. The derivative of a sine wave is a *cosine wave*. This wave has the same shape as the sine wave, but the *phase* is different by 1/4 of a cycle. The term *sinusoid* is sometimes used to describe either a sine wave or a cosine wave.

PROBLEM 3-1

As shown, the peak voltage of the waveform in Fig. 3-1 is ±1 V. Based on this information, what is the peak voltage of its derivative, the waveform shown in Fig. 3-2?

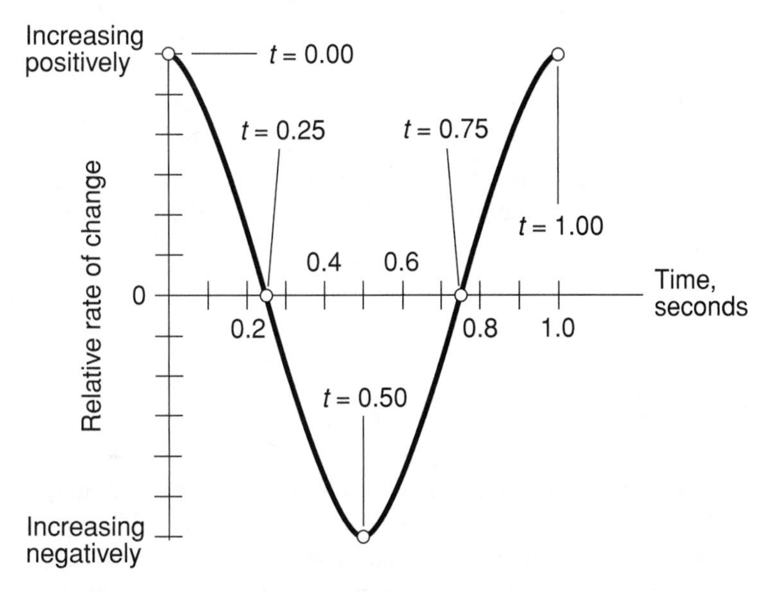

Figure 3-2 A sine wave representing the rate of change in the instantaneous voltage of the wave shown in Figure 3-1.

SOLUTION 3-1

In pure mathematics, the derivative of the sine function is the *cosine function* (symbolized as *cos* in equations). In that context, the peak voltage of the cosine wave, representing the function $y = \cos x$, is ± 1 V, the same as that for the sine wave. But in practical electronics, things aren't so simple. The instantaneous rate of change depends on the frequency of the sine wave, as well as on its peak voltage. Unless we know the angular frequency of a sine wave in radians per second, we cannot say anything about the peak voltage of the derivative of a sine wave.

PROBLEM 3-2

What is the derivative of the derivative, also called the *second derivative*, of a sine wave?

SOLUTION 3-2

This is the derivative of the cosine wave. It turns out to be a negative sine wave, because the derivative of the cosine function is the negative of the sine function. This means that, if a sine wave is *differentiated* (its derivative is taken) twice, the output is 180° out of phase with the input. The peak voltage, however, is uncertain, for the same reason as is the case with the derivative of the original sine wave.

PROBLEM 3-3

What happens to the frequency of a sinusoid when it is differentiated? How is the concentration of energy affected?

SOLUTION 3-3

Differentiation changes the phase of a sinusoid by 90°, and can also change the instantaneous amplitude. But differentiation never changes the frequency of a sinusoid, nor does it change the fact that all the energy in the wave is concentrated at a single frequency.

Circles and Vectors

A sinusoid represents the most efficient possible way that an electrical quantity can "alternate." All the energy is concentrated at a single frequency. If a pure sinusoidal audio-frequency (AF) wave is fed to a headset or a speaker, the resulting sound is a *pure musical tone*. No musical instrument produces a perfect sinusoid, although a flute comes close. Electronic music synthesizers can, however, generate such waves.

THE CIRCULAR MOTION MODEL

The nature of a sine wave can be demonstrated by comparing it with circular motion. Suppose you swing a ball around and around at the end of a string, at a rate of one revolution per second (1 rps). The ball describes a circle in space (Fig. 3-3A). If a friend stands some distance away, with his or her eyes in the plane of the ball's path, your friend sees the ball oscillating back and forth (Fig. 3-3B) with a frequency of 1 Hz. That is one complete cycle per second, because you swing the ball around at 1 rps.

If you graph the position of the ball (as seen by your friend) with respect to time, the result is a sine wave, as shown in Fig. 3-4. This wave has the same fundamental shape as all sinusoids. Some sinusoids are taller than others, and some are stretched out horizontally more than others. But the general waveform is the same in every case. By multiplying or dividing the amplitude and the wavelength of any sinusoid, and by adjusting the phase, it can be made to fit exactly along the curve of any other sinusoid. The *standard sine wave* is the graphical representation of the function $y = \sin x$ in a rectangular coordinate plane.

You might whirl the ball around faster or slower than 1 rps. The string might be made longer or shorter. This would alter the height and/or the frequency of the wave graphed in Fig. 3-4. But a sinusoid can always be reduced to the equivalent of constant, smooth motion in a circular orbit. This is known as the *circular motion model* of a sinusoid.

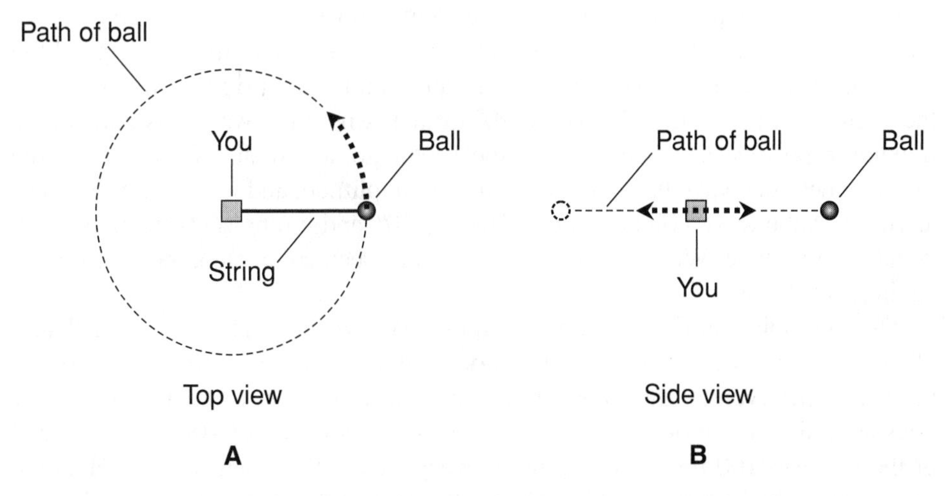

Figure 3-3 A ball can be swung around on a string, and will produce different effects as seen from above (A) and from an edgewise perspective (B).

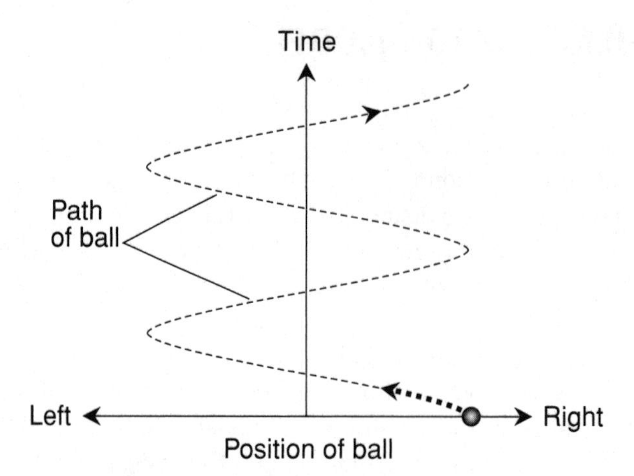

Figure 3-4 Position of ball (horizontal axis) as seen from the side, graphed as a function of time (vertical axis).

ROTATING VECTORS

In the previous chapter, *degrees of phase* were discussed. If you wondered then why phase is spoken of in terms of angular measure, the reason should be clearer now. A circle has 360 angular degrees. A sine wave can be represented as circular motion. Points along a sine wave thus correspond to angles, or positions, around a circle.

Figure 3-5 shows the way a rotating *vector* can be used to represent a sine wave. A vector is a quantity with two independent properties, called *magnitude* (or amplitude) and *direction*. At A, the vector points "east," and this is assigned the value of 0°, where the wave amplitude is zero and is increasing positively. At B, the vector points "north"; this is the 90° instant, where the wave has attained its maximum positive amplitude. At C, the vector points "west." This is 180°, the instant where the wave has gone back to zero amplitude, and is getting more negative. At D, the wave points "south." This is 270°, and it represents the maximum negative amplitude. When a full circle (360°) has been completed, the vector once again points "east."

The four points in Fig. 3-5 are shown on a sine wave graph in Fig. 3-6. Think of the vector as going around counterclockwise at a rate that corresponds to one revolution per cycle of the wave. If the wave has a frequency of 1 Hz, the vector goes around at a rate of 1 rps. If the wave has a frequency of 100 Hz, the speed of the vector is 100 rps, or a revolution every 0.01 s. If the wave is 1 MHz, then the speed of the vector is 1,000,000 rps (10^6 rps), and it goes once around every 0.000001 s (10^{-6} s).

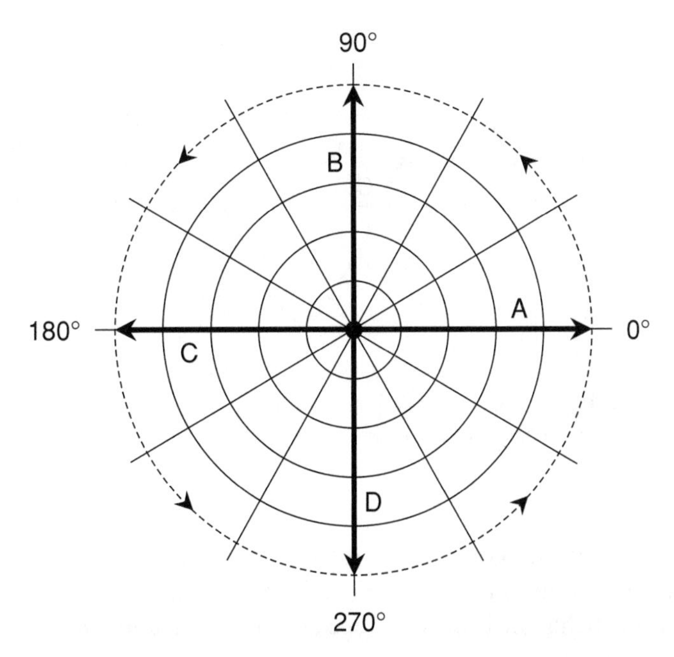

Figure 3-5 Rotating-vector representation of a sine wave. At A, at the start of the cycle; at B, 1/4 of the way through the cycle; at C, halfway through the cycle; at D, 3/4 of the way through the cycle.

The peak amplitude of a pure AC sine wave corresponds to the length of its vector. In Fig. 3-5, time is shown by the angle counterclockwise from "due east." Amplitude is independent of time. The vector length never changes, but its direction does.

PROBLEM 3-4

Suppose the peak amplitude of a sine wave is cut in half, while its frequency is doubled. How will this affect the rotating-vector paradigm?

SOLUTION 3-4

The length of the vector will become half as great, and the number of revolutions per second will double.

PROBLEM 3-5

Suppose the length of the vector in the rotating-vector paradigm depends on its direction. For example, imagine that the vector gets shorter and shorter as its direction progresses from 0° to 180°, so it is half as long at the 180° axis as at the 0° axis.

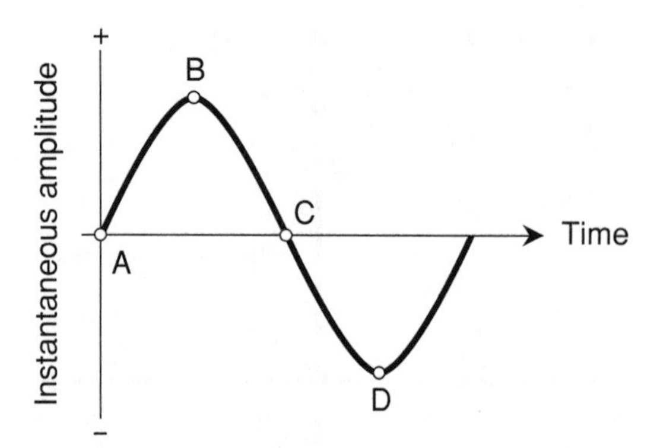

Figure 3-6 The four points for the vector model of Fig. 3-5, shown in a time-domain graph of a sine wave.

Imagine that as the vector continues to rotate the rest of the way around the circle, its length increases, returning to its original length when it gets back to the 0° axis. Suppose that this behavior is exactly repeated from cycle to cycle. How will this affect the waveform?

SOLUTION 3-5

The resulting AC wave will be periodic, because its shape will be the same from cycle to cycle. However, it will not be a sinusoid.

Expressions of Phase Difference

The *phase difference*, also called the *phase angle*, between two waves can have meaning only when those two waves have identical frequencies. If the frequencies differ, even by just a little bit, the relative phase constantly changes, and it's impossible to specify a value for it. In the following discussions of phase angle, therefore, we must assume that the two waves always have identical frequencies.

PHASE COINCIDENCE

The term *phase coincidence* means that two waves begin at exactly the same moment. They are "lined up." This is shown in Fig. 3-7 for two sine waves having different amplitudes. The phase difference in this case is 0°. You could say it's

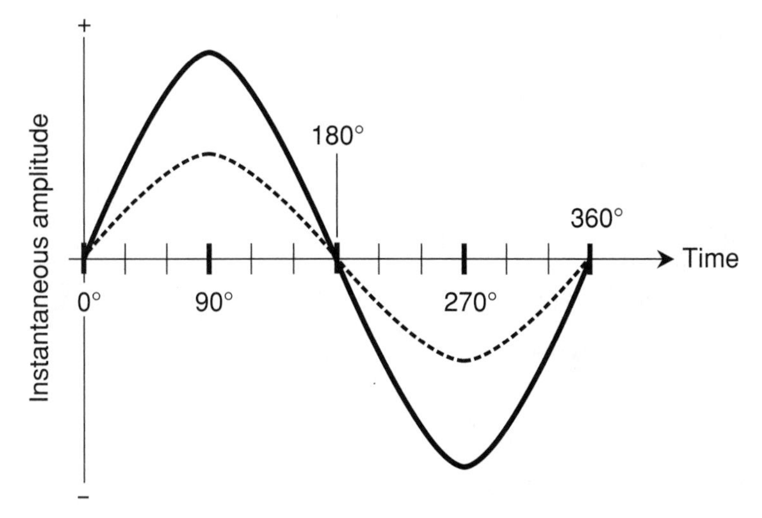

Figure 3-7 Two sine waves in phase coincidence.

some whole-number multiple of 360°, too, but engineers and technicians rarely speak of any phase angle of less than 0° or more than 360°.

If two sine waves are in phase coincidence, and if neither wave has DC superimposed, then the resultant is a sine wave with positive or negative peak amplitudes equal to the sum of the positive and negative peak amplitudes of the composite waves. The phase of the resultant is the same as that of the composite waves.

PHASE OPPOSITION

When two sine waves begin exactly 1/2 cycle, or 180°, apart, they are said to be in *phase opposition*. An example of this is illustrated in Fig. 3-8. In this situation, engineers sometimes say that the waves are *out of phase*, although this expression is a little nebulous because it could be taken to mean some phase difference other than 180°.

If two sine waves have the same amplitudes and are in phase opposition, they cancel each other out. This is because the instantaneous amplitudes of the two waves are equal and opposite at every moment in time.

If two sine waves are in phase opposition, and if neither wave has DC superimposed, then the resultant is a sine wave with positive or negative peak amplitudes equal to the difference between the positive and negative peak amplitudes of the composite waves. The phase of the resultant is the same as the phase of the stronger of the two composite waves.

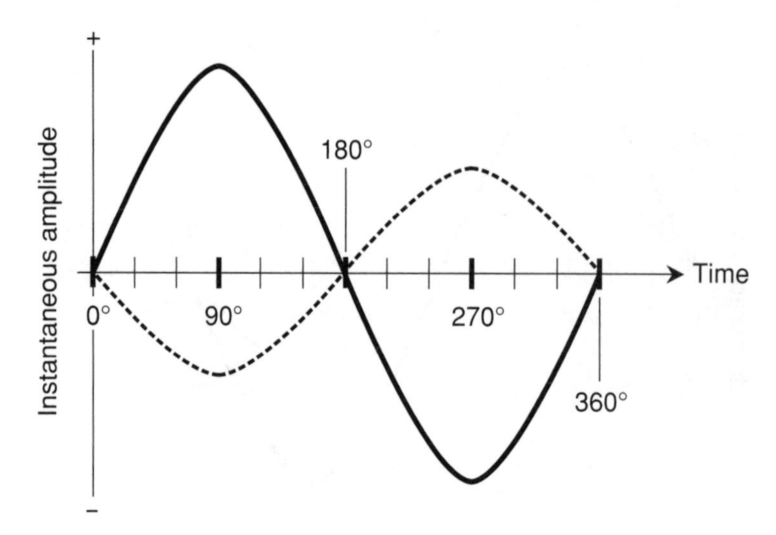

Figure 3-8 Two sine waves in phase opposition.

Any sine wave without superimposed DC has the unique property that, if its phase is shifted by 180°, the resultant wave is the same as turning the original wave "upside-down." Not all waveforms have this property. Perfect square waves do, but some rectangular and sawtooth waves don't, and irregular waveforms almost never do.

INTERMEDIATE PHASE DIFFERENCES

Two sine waves can differ in phase by any amount from 0° (phase coincidence), through 90° (*phase quadrature*, meaning a difference a quarter of a cycle), 180° (phase opposition), 270° (phase quadrature again), to 360° (phase coincidence again).

LEADING PHASE

Imagine two sine waves, called wave *X* and wave *Y*, with identical frequency. If wave *X* begins a fraction of a cycle *earlier* than wave *Y*, then wave *X* is said to be *leading* wave *Y* in phase. For this to be true, *X* must begin its cycle less than 180° before *Y*. Figure 3-9 shows wave *X* leading wave *Y* by 90°. If wave *X* (the dashed line) leads wave *Y* (the solid line), then wave *X* is displaced to the *left* of wave *Y*. In a time-domain graph or display, displacement to the left represents earlier moments in time, and displacement to the right represents later moments in time.

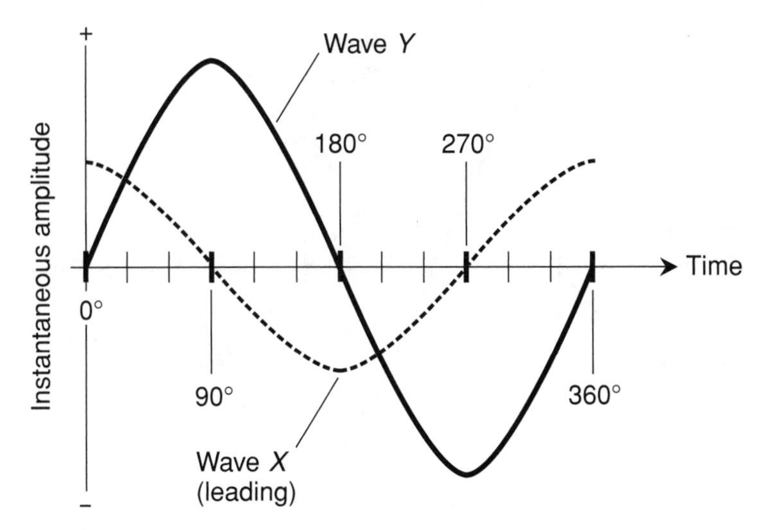

Figure 3-9 Wave X leads wave Y by 90° of phase (1/4 of a cycle).

LAGGING PHASE

Suppose a sine wave X begins its cycle more than 0°, but less than 180°, behind another sine wave Y. Then wave X is *lagging* wave Y. Figure 3-10 shows wave X lagging wave Y by 90°. If wave X (the dashed line) lags wave Y (the solid line), then wave X is displaced to the *right* of wave Y.

PROBLEM 3-6

Suppose the two waves shown in Fig. 3-9 are combined. Will the resultant wave be AC? Will it be periodic? Will it be a sinusoid?

SOLUTION 3-6

The resultant wave will be AC, because its polarity will reverse repeatedly. It will be periodic, because the waveform will be duplicated with each cycle. However, it will not be a sinusoid. If you want to know what this resultant waveform will look like, plot points at phase intervals of approximately 10°, representing the sum of the instantaneous amplitudes of the two waveforms in each case, and then connect the points with a smooth curve.

PROBLEM 3-7

Suppose you have two sinusoid signals X and Y having identical frequency. Imagine that you can adjust the phase difference to any value you want, so wave

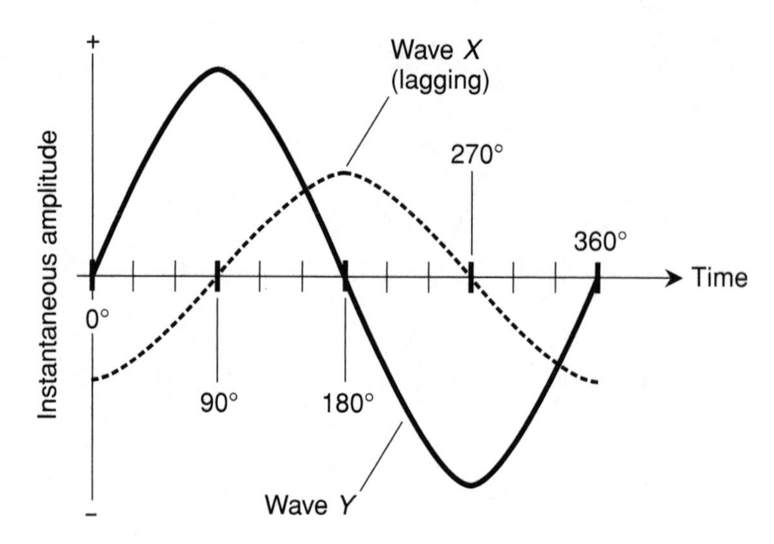

Figure 3-10 Wave X lags wave Y by 90° of phase (1/4 of a cycle).

X leads or lags wave Y by up to 180°. If you want the resultant (composite) wave to be a sinusoid, what must you do?

SOLUTION 3-7

The combination of two sinusoid signals, having the same frequency, will be a sinusoid *if and only if* the two signals are exactly in phase coincidence or exactly in phase opposition. In these special cases, it doesn't matter what the peak amplitudes happen to be, although if they are equal and the waves are in phase opposition, the composite will be no signal at all. If the phase difference is anything other than 0° or 180°, then the resultant will not be a sinusoid, although it will be a periodic AC wave.

Vector Diagrams of Phase Difference

The vector renditions of sine waves, such as those shown in Fig. 3-5, are well suited to showing phase relationships.

ANGULAR DISPLACEMENT

If a sine wave X leads a sine wave Y by some angle q in degrees (where q is greater than 0 but less than 180), then the two waves can be drawn as vectors **X** and **Y**,

with **X** displaced by *q* degrees *counterclockwise* from **Y**. If a sine wave *X* lags a sine wave *Y* by *q* degrees, then **X** appears to point in a direction *clockwise* from **Y** by *q* degrees. If two waves are in phase coincidence, then their vectors point in the same direction. If two waves are in phase opposition, then their vectors point in opposite directions.

FOUR EXAMPLES

The drawings of Fig. 3-11 show four phase relationships between two sine waves *X* and *Y*. At A, *X* is in phase with *Y*. At B, *X* leads *Y* by 90°. At C, *X* and *Y* are 180° apart in phase. At D, *X* lags *Y* by 90°. In all of these examples, think of the vectors

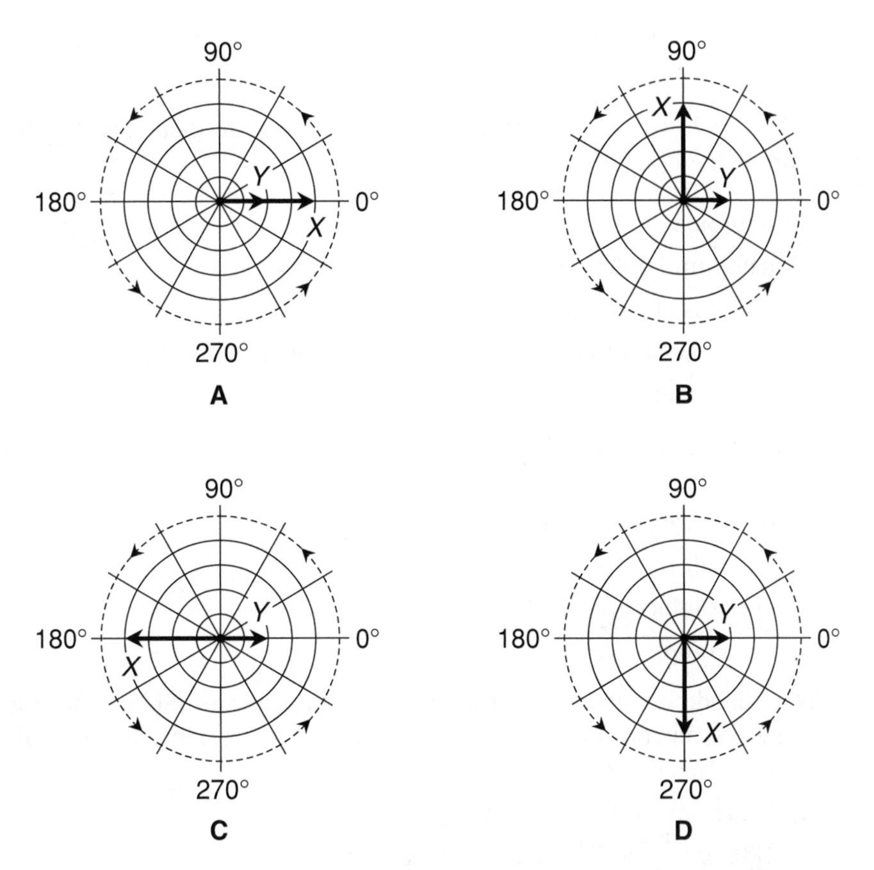

Figure 3-11 Vector representations of phase difference. At A, waves *X* and *Y* are in phase. At B, *X* leads *Y* by 90°. At C, *X* and *Y* are 180° out of phase. At D, *X* lags *Y* by 90°. Time is represented by steady counterclockwise rotation of both vectors.

rotating *counterclockwise* as time passes, but always maintaining the same angle with respect to each other, and always maintaining the same lengths. If the frequency in hertz is f, then the pair of vectors rotates together, counterclockwise, at an angular speed of f, expressed in complete 360° rotations per second (rps).

PROBLEM 3-8

Imagine the situation in any one of the drawings of Fig. 3-11, representing a particular instant in time. Suppose that the two vectors, **X** and **Y**, rotate at different rates. What can be said about the relative phase of the two waves, X and Y, in this case?

SOLUTION 3-8

Their relative phase cannot be defined because their frequencies are not the same.

Quiz

Refer to the text in this chapter if necessary. A good score is at least 8 correct. Answers are in the back of the book.

1. All sinusoids
   (a) have similar general appearance.
   (b) have instantaneous rise and fall times.
   (c) are in the same phase as theoretical sine or cosine waves.
   (d) rise instantly, but decay slowly.
2. A phase difference of 180° in the circular motion model of a sine wave represents
   (a) one-quarter of a rotation.
   (b) one-half of a rotation.
   (c) one complete rotation.
   (d) two complete rotations.
3. You can add or subtract a certain number of degrees of phase to or from a sine wave, and end up with an inverted (upside-down) representation of the original. This number is
   (a) 90, or any odd-number multiple of it.
   (b) 180, or any odd-number multiple of it.
   (c) 270, or any odd-number multiple of it.
   (d) 360, or any odd-number multiple of it.

4. If a wave has a frequency of 440 Hz, how long does it take for 10° of a cycle to occur?
   (a) 0.00273 s.
   (b) 0.000273 s.
   (c) 0.0000631 s.
   (d) 0.00000631 s.
5. Suppose two sinusoidal waves are in phase coincidence. One has peak values of ±3 V and the other has peak values of ±5 V. The resultant has voltages of
   (a) ±8 V pk, in phase with the composites.
   (b) ±2 V pk, in phase with the composites.
   (c) ±8 V pk, in phase opposition with respect to the composites.
   (d) ±2 V pk, in phase opposition with respect to the composites.
6. When a sine wave lags another sine wave by 90°, then the two waves
   (a) are in phase opposition.
   (b) are in phase quadrature.
   (c) differ in amplitude by a factor of 4:1.
   (d) differ in angular frequency by a quarter of a cycle.
7. If wave X leads wave Y by 45°, then
   (a) wave Y is 1/4 of a cycle ahead of wave X.
   (b) wave Y is 1/4 of a cycle behind wave X.
   (c) wave Y is 1/8 of a cycle behind wave X.
   (d) wave Y is 1/16 of a cycle ahead of wave X.
8. Refer to Fig. 3-12. In this example,
   (a) X lags Y by 45°.
   (b) X leads Y by 45°.
   (c) X lags Y by 135°.
   (d) X leads Y by 135°.
9. Suppose Fig. 3-13 represents two sinusoids having identical frequency. Which drawing shows wave X lagging wave Y by 45°?
   (a) Drawing A.
   (b) Drawing B.
   (c) Drawing C.
   (d) Drawing D.

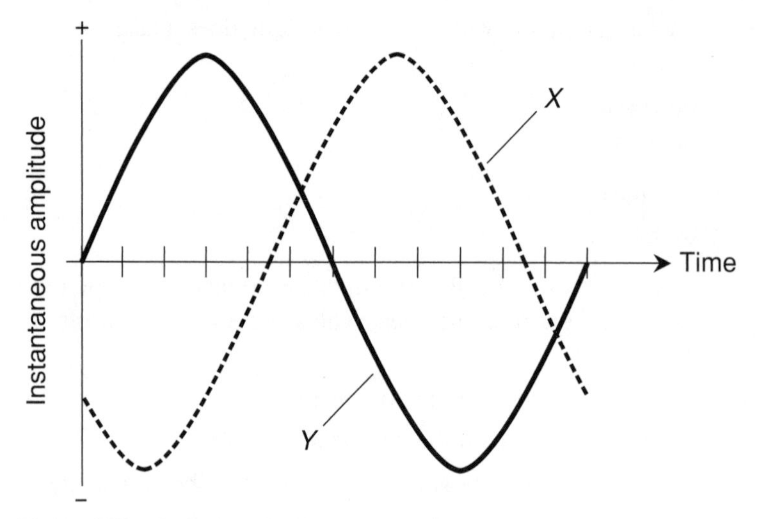

Figure 3-12 Illustration for Quiz Question 8.

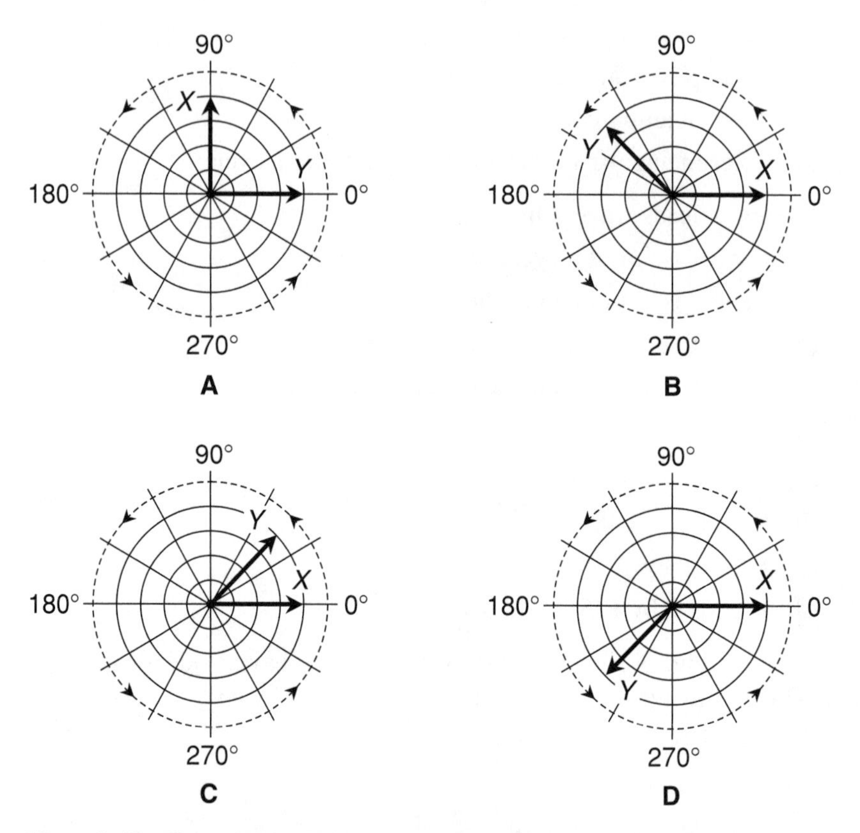

Figure 3-13 Illustration for Quiz Questions 9 and 10.

10. In vector diagrams such as those of Fig. 3-13, the distance from the center of the graph represents
    (a) average amplitude.
    (b) frequency.
    (c) phase.
    (d) peak amplitude.

CHAPTER 4

The Bipolar Transistor

The word *transistor* is a contraction of "current-*trans*ferring res*istor*." Transistors are made from *semiconductor materials*. Transistors are responsible for amplifying audio signals, as well as for generating electronic music.

Semiconductor Materials

Various elements, compounds, and mixtures can function as semiconductors. The two most common materials are *silicon* and a compound of gallium and arsenic known as *gallium arsenide* (often abbreviated GaAs). In the early years of semiconductor technology, *germanium* formed the basis for many semiconductors. Today it is seen occasionally, but not often. Other substances that work as semiconductors are *selenium*, *cadmium* compounds, *indium* compounds, and the oxides of certain metals. For a semiconductor material to have the properties necessary in order to function as electronic components, *impurities* are added. The addition of an impurity to a semiconductor is called *doping*, and such an impurity material is sometimes called a *dopant*.

DONOR IMPURITIES

When an impurity contains an excess of electrons, it is called a *donor impurity*. Adding such a substance causes electrical conduction to take place mainly by electron flow, as in an ordinary metal such as copper. The excess electrons are passed from atom to atom when a voltage exists across the material. Elements that can work as donor impurities include antimony, arsenic, bismuth, and phosphorus. A material containing a donor impurity is called an *N-type semiconductor*, because an electron has a unit negative (N) charge.

ACCEPTOR IMPURITIES

If an impurity has a deficiency of electrons, the dopant is called an *acceptor impurity*. When a substance such as aluminum, boron, gallium, or indium is added to a semiconductor, the material conducts by means of *hole flow*. A *hole* is a spot in an atom where an electron would normally be, but isn't! It's a unit electron deficiency. A semiconductor doped with an acceptor impurity is called a *P-type semiconductor*, because a hole has a unit positive (P) charge.

MAJORITY AND MINORITY CARRIERS

Sometimes electrons account for most of the current in a semiconductor. This is the case if the material has donor impurities, so it is N-type. In other cases, holes account for most of the current. This happens when the material has acceptor impurities, making it P-type. The more abundant charge carriers (either electrons or holes) are called the *majority carriers*. The less abundant ones are called the *minority carriers*. The ratio of majority to minority carriers can vary, depending on the way in which the semiconductor material has been manufactured.

The P-N Junction

When samples of P-type and N-type material are placed in contact with each other, the boundary between them, called the *P-N junction*, behaves in peculiar and useful ways.

FORWARD BIAS

A *semiconductor diode* is formed by joining a piece of P-type material to a piece of N-type material. When the N-type material is negative with respect to the P-type, electrons flow easily from N to P. The N-type semiconductor, which already has an excess of electrons, receives more; the P-type semiconductor, with a shortage of electrons, has some more taken away. The N-type material constantly feeds electrons to the P-type in an "attempt" to create an electron balance, and the battery or power supply keeps "robbing" electrons from the P-type material. This condition is known as *forward bias*. Current flows through the diode easily under these circumstances.

FORWARD BREAKOVER

It takes a specific, well-defined minimum applied voltage for conduction to occur through a semiconductor diode. This is called the *forward breaker voltage*. Depending on the type of material, the forward breakover voltage varies from about 0.3 V to 1 V. If the voltage across the junction is not at least as great as the forward breaker voltage, the diode will not conduct. This effect, known as the *forward breakover effect* or the *P-N junction threshold effect*, can be of use in circuits designed to limit the positive and/or negative peak voltages that signals can attain. The effect can also be used in a device called a *threshold detector*, in which a signal must be stronger than a certain amplitude in order to pass through.

REVERSE BIAS

When the battery or DC power-supply polarity is switched so the N-type material is positive with respect to the P-type, the situation is called *reverse bias*. Electrons in the N-type material are pulled towards the positive charge pole, away from the P-N junction. In the P-type material, holes are pulled toward the negative charge pole, also away from the P-N junction. The zone near the P-N junction, where majority carriers are deficient, is called the *depletion region*. A shortage of majority carriers in a semiconductor means that it cannot conduct well. Thus, the depletion region acts like an electrical insulator. This is why a semiconductor diode will not normally conduct when it is reverse-biased.

AVALANCHE EFFECT

The greater the reverse bias voltage, the more like an electrical insulator a P-N junction gets, up to a point. But if the reverse bias rises past a specific critical value, the voltage overcomes the ability of the junction to prevent the flow of current, and the junction conducts as if it were forward-biased. This phenomenon is called the *avalanche effect* because conduction occurs in a sudden and massive way. Avalanche effect does not damage a P-N junction unless the voltage is extreme. When the voltage drops back below the critical value, the junction behaves normally again.

NPN VERSUS PNP

In a *bipolar transistor*, there are three layers of semiconducting material of alternate types, so there are two P-N junctions. There are two ways in which a bipolar transistor can be manufactured: a P-type layer can be sandwiched between two N-type layers, or an N-type layer can be sandwiched between two P-type layers.

A simplified drawing of an *NPN bipolar transistor* is shown in Fig. 4-1A, and the schematic symbol is shown in Fig. 4-1B. The P type, or center, layer is called the *base*. One of the N-type semiconductor layers is the *emitter,* and the other is the *collector*. Sometimes these are labeled *B*, *E*, and *C* in schematic diagrams. A *PNP bipolar transistor* has two P-type layers, one on either side of a thin, N-type layer (Fig. 4-2A). The schematic symbol is shown in Fig. 4-2B. It's easy to tell whether a bipolar transistor in a diagram is NPN or PNP. If the device is NPN, the arrow at the emitter points outward. If the device is PNP, the arrow at the emitter points inward.

In most applications, an NPN device can be replaced with a PNP device or vice versa, the power-supply polarity can be reversed, and the circuit will work in the same way, provided the new device has the appropriate specifications.

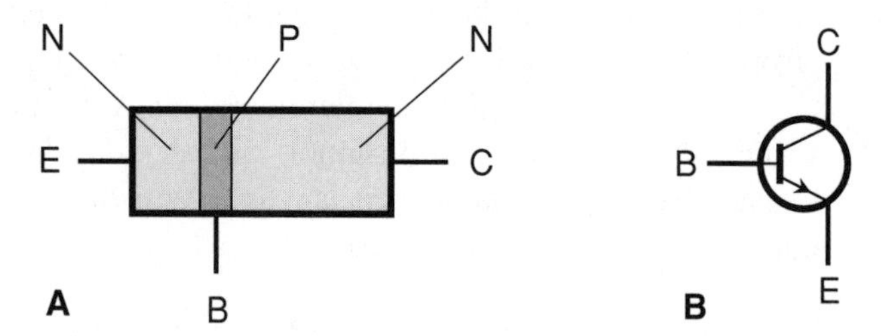

Figure 4-1 At A, pictorial diagram of an NPN transistor. At B, the schematic symbol. Electrodes are E = emitter, B = base, and C = collector.

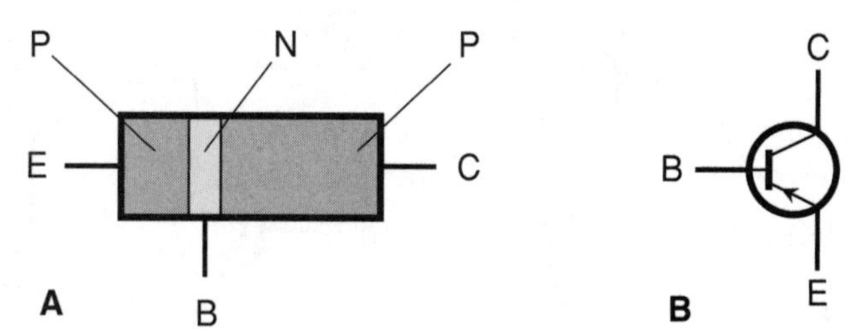

Figure 4-2 At A, pictorial diagram of a PNP transistor. At B, the schematic symbol. Electrodes are E = emitter, B = base, and C = collector.

PROBLEM 4-1

How can the movement of holes produce an electrical current? It seems counter-intuitive. If holes are electron absences, doesn't this mean that a substance with a lot of holes would conduct poorly or not at all?

SOLUTION 4-1

Holes are not physical particles like electrons. They're abstractions! In so-called *hole flow*, electrons actually travel from atom to atom (just as is the case in ordinary electron flow), and the holes "travel" in a direction opposite to the direction in which the electrons go. Electrons move away from negative charge poles and toward positive poles, and holes thus "move" away from positive charge poles and toward negative poles. It is convenient to think of a hole as a real particle in theoretical discussions of current flow in semiconductor materials. But a hole cannot exist all by itself in space, the way an electron can. When you hear or read about holes in technical presentations or papers, keep this in mind.

Biasing

Imagine a device made from two diodes connected in *reverse series* (that is, in series, but in opposite directions). This is good for *modeling* the behavior of bipolar transistors in theoretical discussions.

THE NPN CASE

A dual-diode NPN transistor model is shown in Fig. 4-3A. The base is formed by the connection of the two *anodes*. The emitter is one of the *cathodes*, and the collector

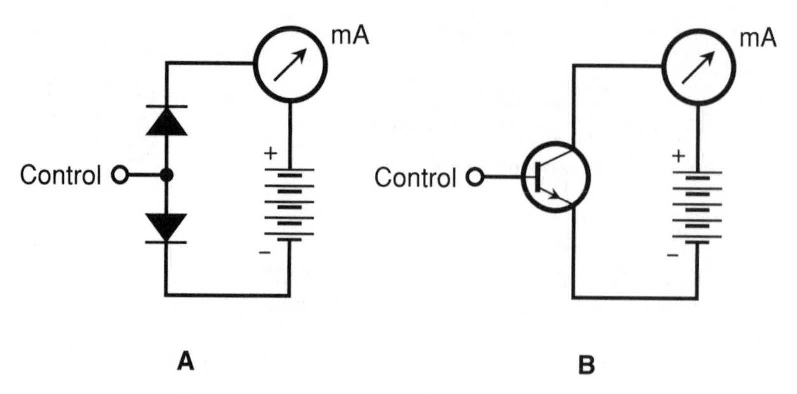

Figure 4-3 At A, the dual-diode model of a simple NPN circuit. At B, the transistor circuit.

is the other cathode. Figure 4-3B shows the equivalent "real-world" NPN transistor circuit. The normal method of biasing an NPN transistor is to have the collector voltage positive with respect to the emitter. This is shown by the connection of the battery in Figs. 4-3A and 4-3B. The normal DC voltages for a transistor power supply range between 3 V and about 50 V. A typical voltage is 12 V. In the model and also in the real-world transistor circuit, the base is labeled "control," because the flow of current through the transistor depends critically on what happens at this electrode.

ZERO BIAS FOR NPN

Suppose the base of a transistor is at the same voltage as the emitter. This is known as *zero bias*. When the bias at the emitter-base (E-B) junction is zero, the emitter-base current, often called simply *base current* and denoted I_B, is zero, and the E-B junction does not conduct. This prevents current from flowing between the emitter and collector, unless a signal is injected at the base to change the situation. Such a signal must, at least momentarily, attain a positive voltage at least equal to the forward breakover voltage of the E-B junction.

REVERSE BIAS FOR NPN

Now imagine that a second battery is connected between the base and the emitter in the circuit of Fig. 4-3B, with the polarity such that E_B becomes negative with respect to the emitter. The addition of this new battery will produce reverse bias at

the E-B junction. No current flows through the E-B junction in this situation, as long as the new battery voltage is not so great that avalanche breakdown occurs. A signal can be injected at the base to cause a flow of current, but such a signal must attain, at least momentarily, a positive voltage high enough to overcome both the reverse bias voltage and the forward breakover voltage of the junction.

FORWARD BIAS FOR NPN

Now suppose that E_B is made positive with respect to the emitter, starting at small voltages and gradually increasing. This is *forward bias*. If the forward bias is less than the forward breakover voltage, no current will flow. But as the base voltage E_B reaches the breakover point, the E-B junction will start to conduct.

The base-collector (B-C) junction of a bipolar transistor is normally reverse-biased. It will remain reverse-biased as long as E_B is less than the supply voltage (in this case 12 V). In practical transistor circuits, it is common for E_B to be positive, and set at a fraction of the supply voltage. Despite the reverse bias of the B-C junction, a significant emitter-collector current, called *collector current* and denoted I_C, will flow once the E-B junction conducts.

In a real transistor circuit such as the one shown in Fig. 4-3B, the meter reading will jump when the forward breakover voltage of the E-B junction is reached. Then a small rise in E_B, attended by a rise in I_B, will cause a large increase in I_C, as shown in Fig. 4-4. If E_B continues to rise, a point will eventually be reached where the I_C-vs.-E_B curve levels off. The transistor will be *saturated* or *in saturation*, conducting as well as it possibly can.

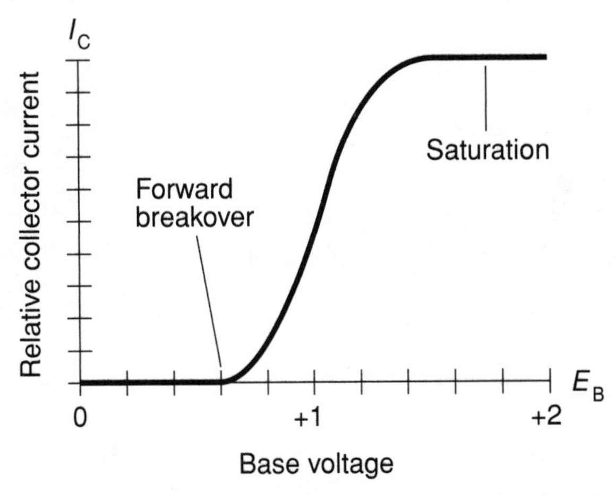

Figure 4-4 Relative collector current (I_C) as a function of base voltage (E_B) for a hypothetical NPN silicon transistor.

THE PNP CASE

For a PNP transistor, the situation is a mirror image of the case for an NPN device. The diodes are reversed, the arrow points inward rather than outward in the transistor symbol, and all the polarities are reversed. The dual-diode PNP model, along with the real-world transistor circuit, are shown in Figs. 4-5 A and 4-5 B. In the discussion above, replace every occurrence of the word "positive" with the word "negative." When the E-B bias is negative and is set at the right level, small changes in E_B cause small changes in I_B, which in turn produce large fluctuations in I_C. If E_B increases negatively, saturation will eventually occur.

PROBLEM 4-2

What is the forward breakover voltage at the E-B junction of a transistor?

SOLUTION 4-2

It depends on the type of semiconductor material. For a silicon transistor, which is the most common type, the forward breakover voltage is approximately 0.6 V.

PROBLEM 4-3

Suppose the base of a bipolar transistor is at the same voltage as the emitter (zero bias), and an AC input signal is applied to the base. For current to flow in the transistor, it is necessary for the instantaneous voltage of the input signal to overcome the forward breakover voltage of the E-B junction. In a silicon transistor, the peak input signal voltage must be at least 0.6 V (positive if the transistor is NPN, and negative if the transistor is PNP). Suppose you do not want the transistor to con-

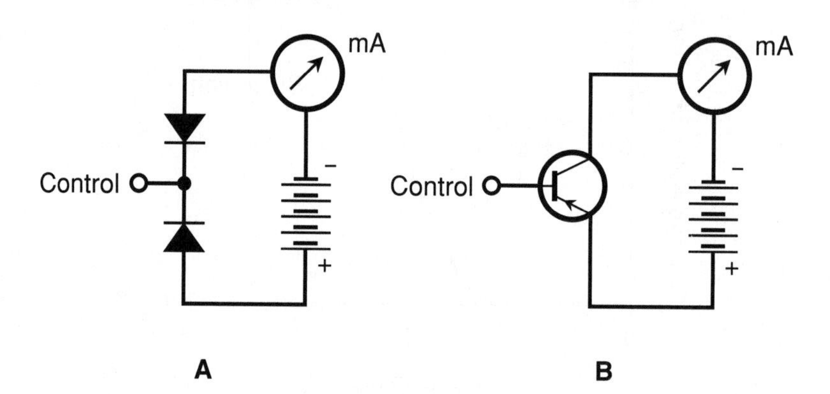

Figure 4-5 At A, the dual-diode model of a simple PNP circuit. At B, the actual transistor circuit.

duct in the absence of an input signal, but you want it to begin conducting when the signal voltage reaches some critical value less than 0.6 V. What can be done to make this happen?

SOLUTION 4-3

The E-B junction can be forward biased at a voltage x, such that x is larger than 0 V but less than 0.6 V. The transistor will then conduct in the presence of an input signal whose peak amplitude is equal to $(0.6 - x)$ V.

Amplification

In a bipolar transistor, a small change in I_B causes a large variation in I_C when the bias is just right. Thus, a transistor can operate as a *current amplifier*. Figure 4-6 is a graph of the collector current as a function of the base current (I_C vs. I_B) for a hypothetical transistor. From this graph, you can see that there are some bias values at which a transistor won't provide any current amplification. If the transistor is in saturation, the I_C-vs.-I_B curve is horizontal. A small change in I_B, in these portions of the curve, causes little or no change in I_C. But if the transistor is biased near the middle of the straight-line part of the curve in Fig. 4-6, the transistor will work as a current amplifier.

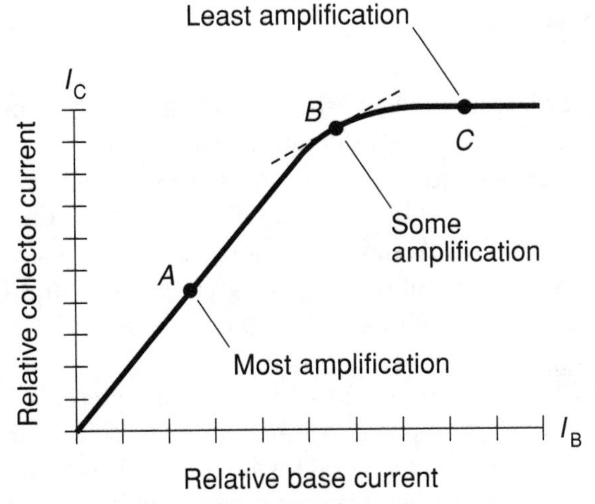

Figure 4-6 Three different transistor bias points. The most amplification is obtained at or near point *A*, where the bias is near the middle of the straight-line portion of the curve.

STATIC CURRENT AMPLIFICATION

Current amplification is often called *beta* by engineers. It can range from a factor of just a few times up to hundreds of times. The beta of a transistor can be expressed as the *static forward current transfer ratio*, abbreviated H_{FE}. Mathematically, this is the collector current divided by the base current:

$$H_{FE} = I_C/I_B$$

For example, if a base current, I_B, of 1 mA results in a collector current, I_C, of 35 mA, then $H_{FE} = 35/1 = 35$. If $I_B = 0.5$ mA and $I_C = 35$ mA, then $H_{FE} = 35/0.5 = 70$. The H_{FE} specification for a particular transistor represents the greatest amount of current amplification that can be obtained with it.

DYNAMIC CURRENT AMPLIFICATION

A more practical way to define current amplification is as the ratio of the difference in I_C to the difference in I_B that occurs when a small signal is applied to the base of a transistor. Abbreviate the words "the difference in" by *d*. Then, according to this second definition:

$$\text{Current amplification} = dI_C/dI_B$$

OPERATING POINT

In Fig. 4-6, three different points are shown, corresponding to three different bias scenarios. The ratio dI_C/dI_B is different for each of the points in this graph. Geometrically, dI_C/dI_B at a given point is the *slope* of a line that is *tangent* to the curve at that point. The tangent line for point *B* in Fig. 4-6 is a dashed, straight line; the tangent lines for points *A* and *C* lie along the curve and are therefore not shown. The steeper the slope of the line, the greater is dI_C/dI_B. Point *A* provides the highest value of dI_C/dI_B, as long as the input signal is not too strong. This value is very close to H_{FE}.

For small-signal amplification, point *A* in Fig. 4-6 represents a good bias level. Engineers would say that it's a good *operating point*. At point *B*, dI_C/dI_B is smaller than at point *A*, so point *B* is not as good for small-signal amplification. At point *C*, dI_C/dI_B is practically zero. The transistor won't amplify much, if at all, if it is biased at this point.

OVERDRIVE

Even when a transistor is biased for the greatest possible current amplification (at or near point *A* in Fig. 4-6), a strong AC input signal can drive it to point *B* or beyond during part of the signal cycle. Then, dI_C/dI_B is reduced, as shown in Fig. 4-7. Points *X* and *Y* in the graph represent the instantaneous current extremes during the signal cycle in this particular case. The slope of the line connecting these points, representing the current amplification, is less than the slope of a line tangent to the optimum operating point *A* in Fig. 4-6.

When conditions are like those in Fig. 4-7, a transistor amplifier will cause *distortion* in the signal. This means that the output wave will not have the same shape as the input wave. This phenomenon is called *nonlinearity*. It can sometimes be tolerated, but often it is undesirable. When the input signal to a transistor amplifier is too strong, the condition is called *overdrive*, and the amplifier is said to be *overdriven*.

Overdrive can cause problems other than signal distortion. An overdriven transistor is in or near saturation during part of the input signal cycle. This reduces circuit efficiency, causes excessive collector current, and can overheat the base-collector (B-C) junction. Sometimes overdrive can destroy a transistor.

PROBLEM 4-4

Suppose that when the base current in a bipolar transistor circuit changes by 10 microamperes (10 μA), the collector current changes by 1 milliampere (1 mA). What is the dynamic current amplification?

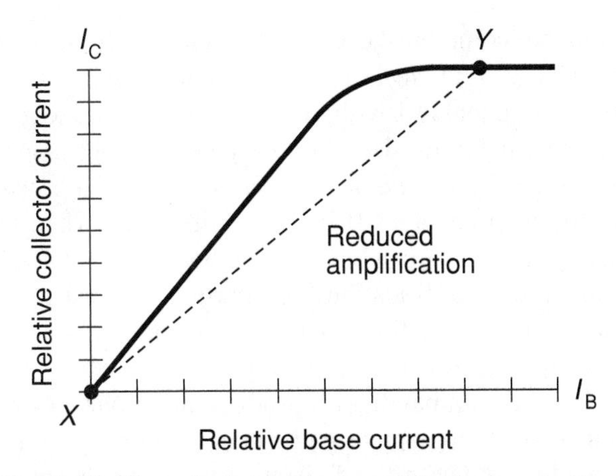

Figure 4-7 Excessive input reduces amplification because it drives a transistor near, or into, saturation during part of the cycle.

SOLUTION 4-4

First, convert the current values to the same units. Let's use microamperes. Therefore, the collector current changes by 1000 μA when the base current changes by 10 μA. That means $dI_C = 1000$ and $dI_B = 10$. Plugging these numbers into the formula for dynamic current amplification, we get this:

$$\begin{aligned}\text{Current amplification} &= dI_C/dI_B \\ &= 1000/10 \\ &= 100\end{aligned}$$

Gain versus Frequency

Another important specification for a transistor is the range of frequencies over which it can be used as an amplifier. All transistors have an *amplification factor*, or *gain*, that decreases as the signal frequency increases. Some devices work well only up to a few megahertz; others can be used to several gigahertz.

Gain can be expressed in various ways. In the above discussion, you learned a little about *current gain*, expressed as a ratio. You will also hear about *voltage gain* or *power gain* in amplifier circuits. These, too, can be expressed as ratios. For example, if the voltage gain of a circuit is 15, then the output signal voltage (rms, peak, or peak-to-peak) is 15 times the input signal voltage. If the power gain of a circuit is 25, then the output signal power is 25 times the input signal power.

Two expressions are commonly used for the gain-vs.-frequency behavior of a bipolar transistor. The *gain bandwidth product*, abbreviated f_T, is the frequency at which the gain becomes equal to 1 with the emitter connected to ground. If you try to make an amplifier using a transistor at a frequency higher than its f_T specification, you are bound to fail. The *alpha cutoff frequency* of a transistor is the frequency at which the gain becomes 0.707 times its value when the input signal frequency is 1 kHz. A transistor can have considerable gain at its alpha cutoff frequency. By looking at this specification for a particular transistor, you can get an idea of how rapidly it loses its ability to amplify as the frequency goes up. Some devices "die off" faster than others.

Figure 4-8 shows the gain bandwidth product and alpha cutoff frequency for a hypothetical transistor, on a graph of gain versus frequency. Note that the scales of this graph are not linear; that is, the divisions are not evenly spaced. This type of graph is called a *log-log* graph because both scales are *logarithmic*.

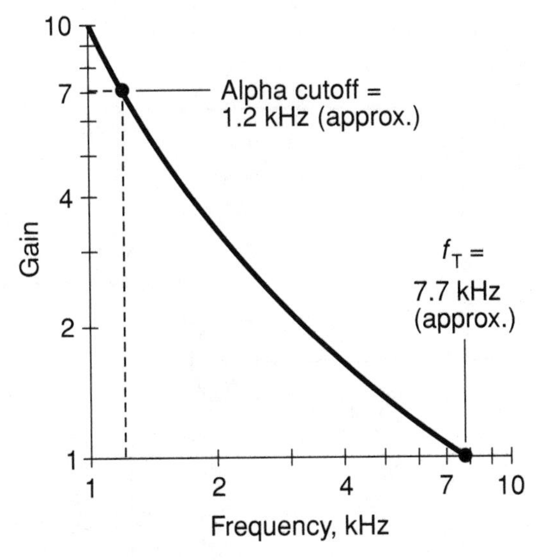

Figure 4-8 Alpha cutoff and gain bandwidth product for a hypothetical transistor.

Common-Emitter Circuit

A transistor can be hooked up in three general ways. The emitter can be grounded for signal, the base can be grounded for signal, or the collector can be grounded for signal. An often-used arrangement is the *common-emitter circuit*. "Common" means "grounded for the signal." The basic configuration is shown in Fig. 4-9.

A terminal can be at ground potential for a signal, and yet have a significant DC voltage. In the circuit shown, capacitor C_1 appears as a short circuit to the AC signal, so the emitter is at *signal ground*. But resistor R_1 causes the emitter to have a certain positive DC voltage with respect to ground (or a negative voltage, if a PNP transistor is used). The exact DC voltage at the emitter depends on the resistance of R_1, and on the bias. The bias is set by the ratio of the values of resistors R_2 and R_3. The bias can be anything from zero, or ground potential, to +12 V, the supply voltage. Normally it is a couple of volts.

Capacitors C_2 and C_3 block DC to or from the input and output circuitry (whatever that might be) while letting the AC signal pass. Resistor R_4 keeps the output signal from being shorted out through the power supply. A signal enters the circuit through C_2, where it causes the base current, I_B, to vary. The small fluctuations in I_B cause large changes in the collector current, I_C. This current passes through

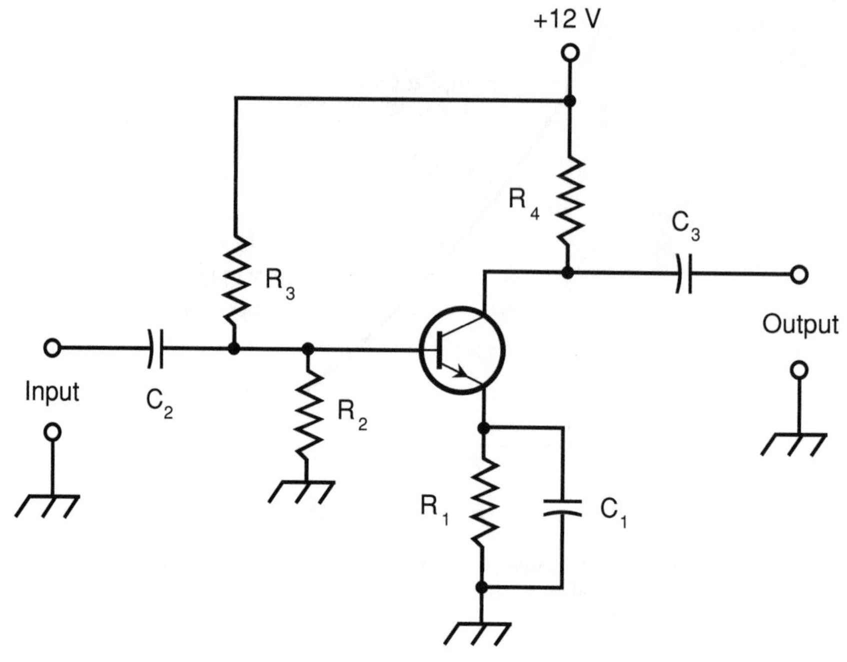

Figure 4-9 Common-emitter configuration. This diagram shows an NPN transistor circuit.

resistor R_4, causing a fluctuating DC voltage to appear across this resistor. The AC part of this passes unhindered through capacitor C_3 to the output.

The circuit of Fig. 4-9 is the basis for many audio amplifiers. The common-emitter configuration is capable of producing the largest signal amplification of any of the three arrangements shown here. For sine-wave signals, the output is 180° out of phase with the input. In general, the output wave is inverted with respect to the input wave.

Common-Base Circuit

As its name implies, the *common-base circuit* (Fig. 4-10) has the base at signal ground. The DC bias is the same as for the common-emitter circuit, but the input signal is applied at the emitter, instead of at the base. This causes fluctuations in the voltage across R_1, causing variations in I_B. The result of these small current fluctuations is a large change in the current through R_4. Therefore, amplification occurs. The output wave is in phase with the input wave.

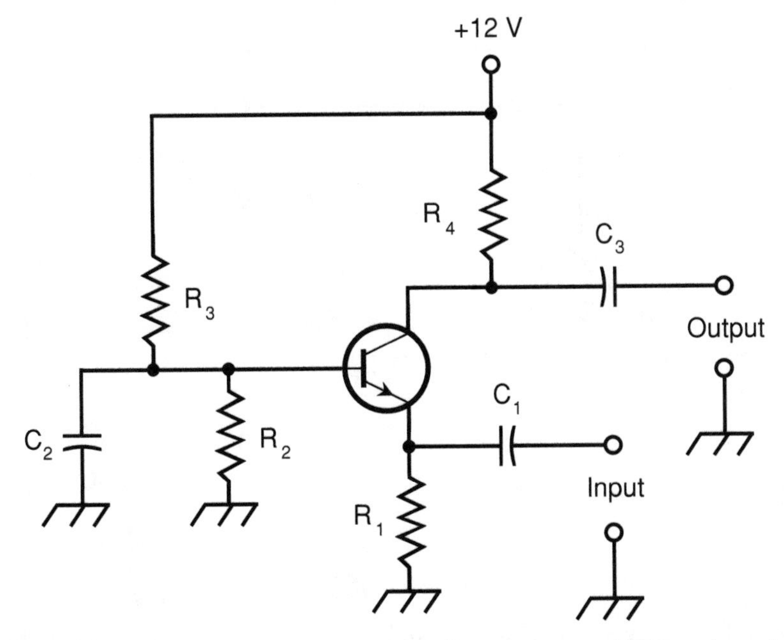

Figure 4-10 Common-base configuration. This diagram shows an NPN transistor circuit.

The signal enters through capacitor C_1. Resistor R_1 keeps the input signal from being shorted to ground. Bias is provided by R_2 and R_3. Capacitor C_2 keeps the base at signal ground. Resistor R_4 keeps the signal from being shorted out through the power supply. The output is taken through C_3.

The common-base circuit provides somewhat less gain than a common-emitter circuit. However, it is more stable than the common-emitter configuration in some applications, especially in power amplifiers. This can be important if you plan to design a high-power home stereo system, or if you want to build a set of amplifiers for use in live music performances.

Common-Collector Circuit

A *common-collector circuit* (Fig. 4-11) operates with the collector at signal ground. This example shows the use of a PNP transistor, in contrast to the previous two examples in which NPN transistors are used. Aside from the transistor symbol, the only difference in this diagram from the equivalent NPN circuit is the power supply polarity.

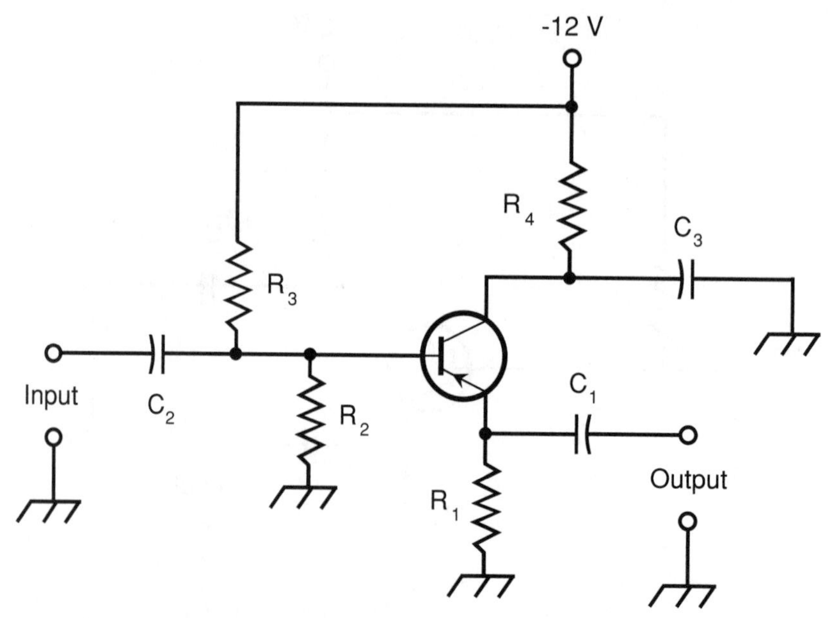

Figure 4-11 Common-collector configuration, also known as an emitter follower. This diagram shows a PNP transistor circuit.

In the common-collector configuration, the input is applied at the base, just as it is with the common-emitter circuit. The signal passes through C_2 onto the base of the transistor. Resistors R_2 and R_3 provide the correct bias for the base. Resistor R_4 limits the current through the transistor. Capacitor C_3 keeps the collector at signal ground. A fluctuating direct current flows through R_1, and a fluctuating DC voltage therefore appears across it. The AC part of this voltage passes through C_1 to the output. Because the output follows the emitter current, this circuit is sometimes called an *emitter follower circuit.*

The output wave of a common-collector circuit is in phase with the input wave. The input impedance is high but the output impedance is low. For this reason, the common-collector circuit can be used to match high impedances to low impedances. When well designed, this type of circuit works over a wide range of frequencies, and is a low-cost alternative to a broadband impedance-matching transformer.

PROBLEM 4-5

In the common-emitter circuit (Fig. 4-9), what would happen if resistor R_1 were to burn out and its resistance become theoretically infinite?

SOLUTION 4-5

In this situation, current could not flow through the transistor, and the collector current would therefore drop to zero. The circuit, in effect deprived of its source of DC power, would lose its ability to amplify.

PROBLEM 4-6

In the common-base circuit (Fig. 4-10), what would happen if resistor R_2 were to burn out and its resistance become theoretically infinite?

SOLUTION 4-6

This would drastically alter the base bias, putting the transistor into a state of saturation or near-saturation. The collector current would rise, and the circuit would lose most, or all, of its ability to amplify.

Quiz

Refer to the text in this chapter if necessary. A good score is at least 8 correct. Answers are in the back of the book.

1. In many cases, a PNP transistor can be replaced with an NPN device and the circuit will do the same thing, provided that
   (a) the power supply or battery polarity is reversed.
   (b) the collector and emitter leads are interchanged.
   (c) the arrow is pointing inward.
   (d) Forget it! A PNP transistor can never be replaced with an NPN transistor.
2. A bipolar transistor has
   (a) three P-N junctions.
   (b) three semiconductor layers.
   (c) two N-type layers around a P-type layer.
   (d) a low avalanche voltage.
3. The current through a transistor depends on
   (a) E_C.
   (b) E_B relative to E_C.
   (c) I_B.
   (d) More than one of the above

4. When a transistor is conducting as much as it can, it is said to be
   (a) in a state of cutoff.
   (b) in a state of saturation.
   (c) in a state of reverse bias.
   (d) in a state of avalanche breakdown.

5. Refer to the curve shown in Fig. 4-12. The best point at which to operate this transistor as a sensitive weak-signal amplifier is
   (a) point *A*.
   (b) point *B*.
   (c) point *C*.
   (d) point *D*.

6. In Fig. 4-12, the forward-breakover point for the E-B junction is nearest to
   (a) no point on this graph.
   (b) point *B*.
   (c) point *C*.
   (d) point *D*.

7. In Fig. 4-12, saturation is nearest to
   (a) point *A*.
   (b) point *B*.
   (c) point *C*.
   (d) point *D*.

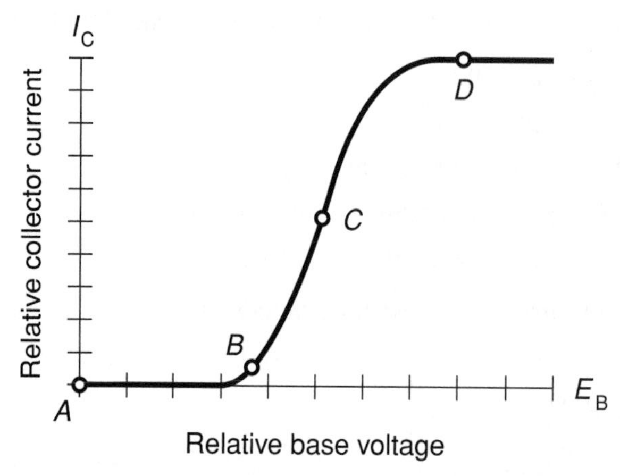

Figure 4-12 Illustration for Quiz Questions 5 through 8.

8. In Fig. 4-12, the greatest gain is achievable at
   (a) point *A*.
   (b) point *B*.
   (c) point *C*.
   (d) point *D*.
9. Which of the following configurations can produce the greatest signal gain under ideal conditions?
   (a) a common-emitter circuit.
   (b) a common-base circuit.
   (c) a common-collector circuit.
   (d) They're all the same in this respect.
10. Suppose that the input signal to a transistor amplifier results in saturation during part of the cycle. This produces
    (a) the least possible amplification.
    (b) signal distortion.
    (c) avalanche effect.
    (d) low output impedance.

The Field-Effect Transistor

The bipolar transistor isn't the only semiconductor transistor that can amplify. The other major category of transistor is the *field-effect transistor* (FET). There are two main types: the *junction FET* (JFET) and the *metal-oxide FET* (MOSFET).

Principle of the JFET

In a JFET, the current varies because of the effects of an electric field within the device. Charge carriers (electrons or holes) flow from the *source* (S) electrode to the *drain* (D) electrode. This results in a *drain current*, I_D, that is normally the same as the *source current*, I_S. The rate of flow of charge carriers—that is, the current—depends on the voltage at a control electrode called the *gate* (G). Fluctuations in *gate voltage*, E_G, cause changes in the current through the *channel*, which is the path between the source and the drain. The current through the channel is normally equal to I_D. Small fluctuations in E_G can cause large variations in I_D. This variable drain current can produce significant fluctuations in the voltage across an output resistance.

N-CHANNEL VERSUS P-CHANNEL

Figure 5-1 is a simplified drawing of an *N-channel JFET* and its schematic symbol. The N-type semiconductor material forms the channel, or the path for charge carriers. The majority carriers are electrons. The drain is placed at a positive DC voltage with respect to the source.

In an N-channel device, the gate consists of P-type material. Another section of P-type material, called the *substrate*, forms a boundary on the side of the channel opposite the gate. The voltage on the gate produces an electric field that interferes with the flow of charge carriers through the channel. The more negative E_G gets, the more the electric field chokes off the current though the channel, and the smaller I_D becomes.

A *P-channel JFET* (Fig. 5-2) has a channel consisting of P-type semiconductor material. The majority charge carriers are holes. The drain is placed at a negative DC voltage with respect to the source. The more positive E_G gets, the more the electric field chokes off the current through the channel, and the smaller I_D becomes.

You can recognize the N-channel JFET in schematic diagrams by the arrow pointing inward at the gate, and the P-channel JFET by the arrow pointing out-

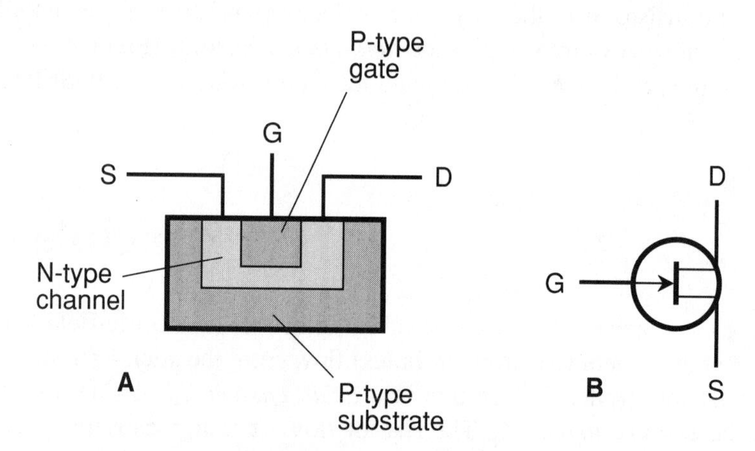

Figure 5-1 At A, a pictorial diagram of an N-channel JFET. At B, the schematic symbol. Electrodes are S = source, G = gate, and D = drain.

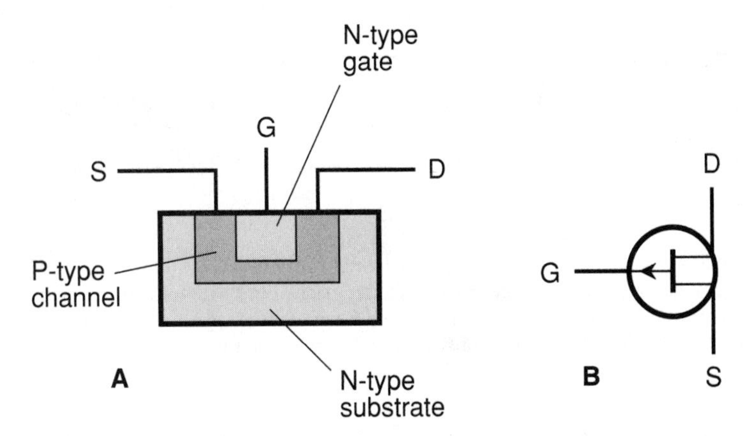

Figure 5-2 At A, a pictorial diagram of a P-channel JFET. At B, the schematic symbol. Electrodes are S = source, G = gate, and D = drain.

ward. Also, you can tell which is which (sometimes arrows are not included in schematic diagrams) by the power-supply polarity. A positive drain indicates an N-channel JFET, and a negative drain indicates a P-channel JFET.

In electronic circuits, N-channel and P-channel devices can do the same kinds of things. The main difference is the polarity. An N-channel device can almost always be replaced with a P-channel JFET, and the power-supply polarity reversed, and the circuit will still work if the new device has the right specifications. Some JFETs work well as low-level amplifiers (such as microphone preamplifiers) or audio oscillators; others are made for power amplification.

Field-effect transistors have certain advantages over bipolar transistors. Perhaps the most important is that FETus, in general, generate less internal noise than bipolar transistors. This can be important in professional audio work. Field-effect transistors have high input impedance. Devices with high input impedance draw almost no power from a signal source, so they are ideal for use in amplifiers that must be highly sensitive, or in amplifiers followed by multiple stages to obtain high audio power with a minimum of amplified electrical noise.

DEPLETION AND PINCHOFF

The JFET works because the voltage at the gate causes an electric field that interferes, more or less, with the flow of charge carriers along the channel. A simplified drawing of the situation for an N-channel device is shown in Fig. 5-3.

As the drain voltage E_{D} increases, so does the drain current I_{D}, up to a certain level-off value. This is true as long as the gate voltage E_{G} is constant, and is not

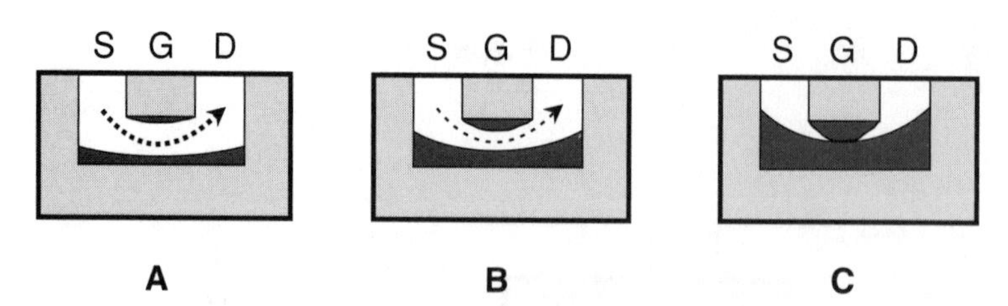

Figure 5-3 At A, the depletion region (darkest area) is narrow, the channel (white area) is wide, and many charge carriers (heavy dashed line) flow. At B, the depletion region is wider, the channel is narrower, and fewer charge carriers (light dashed line) flow. At C, the depletion region obstructs the channel, and no charge carriers flow.

too large negatively. But as E_G becomes increasingly negative (Fig. 5-3A), a *depletion region* (shown as a solid dark area) begins to form in the channel. Charge carriers cannot flow in this region; they must pass through a narrowed channel. The more negative E_G becomes, the wider the depletion region gets, as shown in drawing B. Ultimately, if the gate becomes negative enough, the depletion region completely obstructs the flow of charge carriers. This condition is called *pinchoff*, and is illustrated at C.

JFET BIASING

Two biasing methods for N-channel JFET circuits are shown in Fig. 5-4. In Fig. 5-4A, the gate is grounded through resistor R_2. The source resistor, R_1, limits the current through the JFET. The drain current, I_D, flows through R_3, producing a voltage across this resistor. The AC output signal passes through C_2. In Fig. 5-4B, the gate is connected through potentiometer R_2 to a voltage that is negative with respect to ground. Adjusting this potentiometer results in a variable negative E_G between R_2 and R_3. Resistor R_1 limits the current through the JFET. The drain current, I_D, flows through R_4, producing a voltage across it. The AC output signal passes through C_2.

In both of these circuits, the drain is placed at a positive DC voltage relative to ground. For a P-channel JFET, reverse the polarities in Fig. 5-4. Typical power-supply voltages in JFET circuits are comparable to those for bipolar transistor circuits. The voltage between the source and drain, abbreviated E_D, can range from about 3 V to 150 V DC; most often it is 6 to 12 V DC. The biasing arrangement in Fig. 5-4A is preferred for weak-signal amplifiers, low-level amplifiers, and oscillators. The scheme at B is more often employed in power amplifiers having substantial input signal amplitudes.

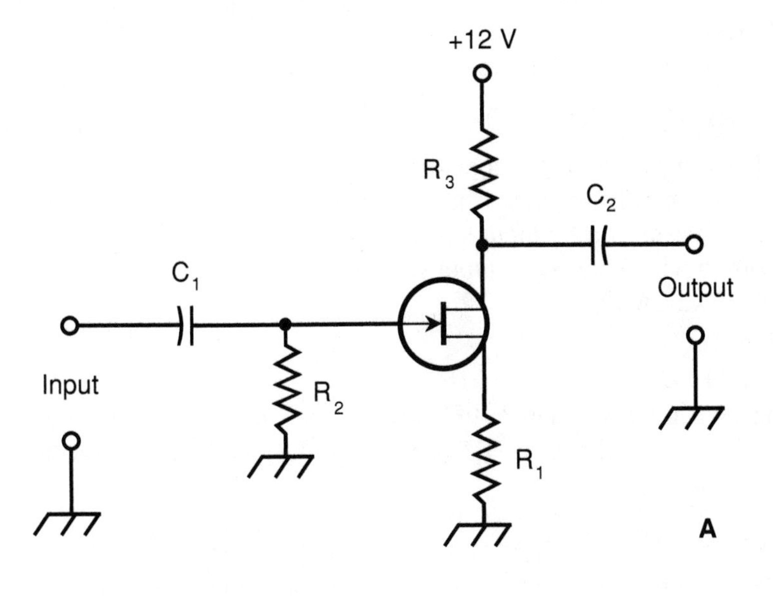

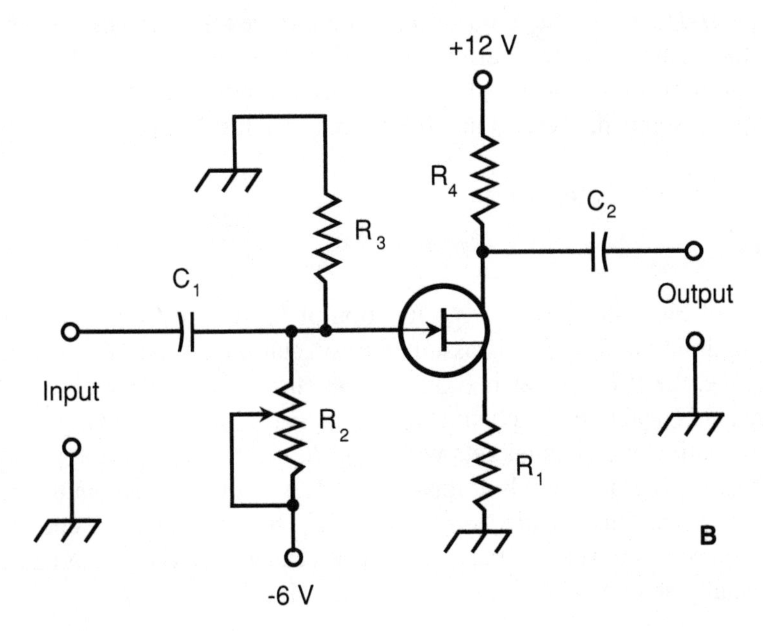

Figure 5-4 Two methods of biasing an N-channel JFET. At A, fixed gate bias; at B, variable gate bias.

PROBLEM 5-1
What happens if a JFET is biased significantly beyond the pinchoff point?

SOLUTION 5-1
When this is done, the input signal must have a certain minimum peak voltage before current can flow in the channel. Weak input signals do not produce any output whatsoever. Strong input signals can be amplified under these conditions, but distortion will occur. The output waveform will not be a faithful reproduction of the input waveform.

PROBLEM 5-2
What happens if an N-channel JFET is biased so the gate is positive with respect to the source, or a P-channel JFET is biased so the gate is negative with respect to the source?

SOLUTION 5-2
Under these conditions, the channel is wide open, and remains that way even in the presence of a weak or moderate input signal. As a result, drain current flows constantly, and there is little or no amplification. If the input signal becomes strong enough, the channel narrows during part of the input signal cycle. This may allow for some amplification, but as in the case of bias beyond pinchoff, distortion occurs. Biasing of this sort is rarely used in a JFET, because it results in inefficient operation.

Amplification

The graph of Fig. 5-5 shows I_D as a function of E_G for a hypothetical N-channel JFET. The drain voltage, E_D, is assumed to be constant. When E_G is fairly large and negative, the JFET is pinched off, and no current flows through the channel. As E_G gets less negative, the channel opens up, and I_D begins flowing. As E_G gets still less negative, the channel gets wider and I_D increases. As E_G approaches the point where the S-G junction is at forward breakover, the channel conducts as well as it possibly can. This is called *saturation*. If E_G becomes positive enough so the S-G junction conducts, some of the current in the channel "leaks out" through the gate. This is usually an undesirable phenomenon in a JFET.

THE FET AMPLIFIES VOLTAGE

The best amplification for weak signals is obtained when E_G is such that the slope of the curve in Fig. 5-5 is the steepest. This is shown roughly by the range marked *X*.

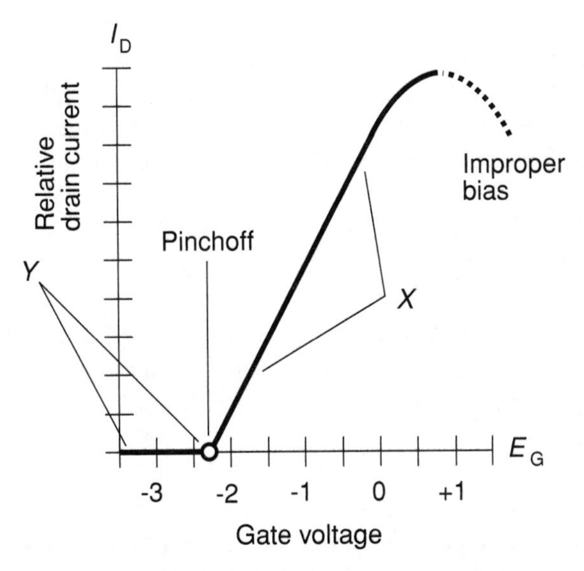

Figure 5-5 Relative drain current (I_D) as a function of gate voltage (E_G) for a hypothetical N-channel JFET.

For power amplification in radio-frequency (RF) amplifiers, results are often best when the JFET is biased at, or beyond, pinchoff, in the range marked *Y*. This biasing scenario is rarely desirable for audio amplification unless the intent is to obtain raw power with no concern about distortion (as, for example, in an electronic alarm or siren).

In either circuit shown in Fig. 5-4, I_D passes through the drain resistor. Small fluctuations in E_G cause large changes in I_D, and these variations in turn produce wide swings in the DC voltage across R_3 (in the circuit at A) or R_4 (in the circuit at B). The AC part of this voltage goes through capacitor C_2, and appears at the output as a signal of much greater AC voltage than that of the input signal at the gate. All of this is based on the assumption that the JFET is biased for weak signal amplification (the range marked X in Fig. 5-5).

DRAIN CURRENT VERSUS DRAIN VOLTAGE

It seems reasonable to suppose that the current I_D, passing through the channel of a JFET, increases in a linear manner with increasing drain voltage E_D, but it is not what happens in most JFETs. Instead, I_D rises for awhile as E_D increases steadily, and then I_D starts to level off. The current I_D can be plotted graphically as a function of E_D for various values of E_G. When this is done, the result is a *family of characteristic curves* for the JFET. The graph of Fig. 5-6 shows a family of characteristic curves for a hypothetical N-channel device. The graph of I_D vs. E_G, one example of which is shown in Fig. 5-5, is also an important specification that engineers consider when choosing a JFET for a particular application.

TRANSCONDUCTANCE

In a bipolar transistor, as you recall from the last chapter, *dynamic current amplification* quantifies the extent to which a bipolar transistor amplifies a signal. The JFET equivalent of this specification is called *dynamic mutual conductance* or *transconductance*.

Refer again to Fig. 5-5. Suppose that E_G is a certain value, resulting in a certain current I_D. If the gate voltage changes by a small amount dE_G, then the drain current changes by a certain increment dI_D. The transconductance is the ratio dI_D/dE_G. Geometrically, this translates to the slope of a line tangent to the curve of Fig. 5-5.

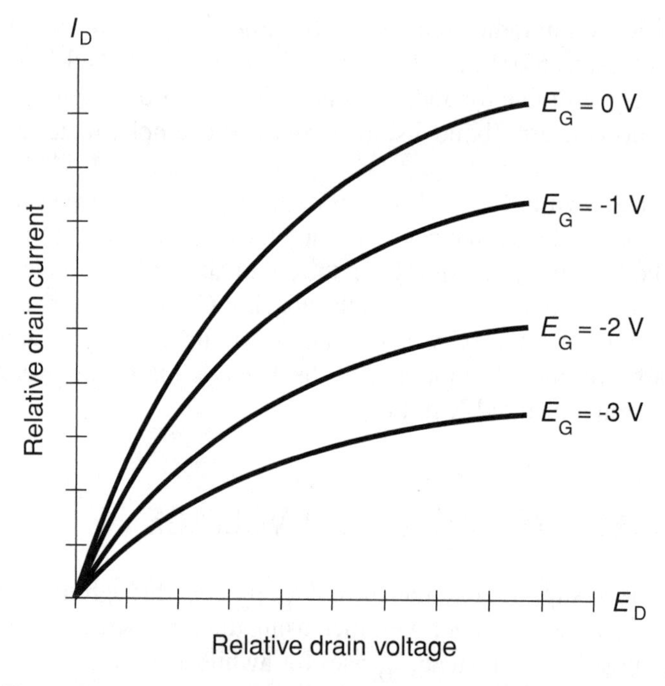

Figure 5-6 A family of characteristic curves for a hypothetical N-channel JFET.

The value of dI_D/dE_G is not the same at every point along the curve. When the JFET is biased beyond pinchoff, in the region marked *Y*, the slope of the curve is zero. Then there is no fluctuation in I_D when E_G changes by small amounts. There can be a change in I_D when there is a change in E_G only when the channel conducts current. The region where the transconductance, dI_D/dE_G, is the greatest is marked *X*, where the slope of the curve is steepest. This portion of the curve represents conditions where the most gain can be obtained from the device. The center of the straight-line portion of the curve is optimum for both gain and linearity.

PROBLEM 5-3

Consider an FET circuit that is biased to a point near the middle of the region marked *X* in Fig. 5-5. Under what conditions can an amplifier produce an output waveform with a minimum of distortion (that is, good linearity) so the output signal is a faithful reproduction of the input signal?

SOLUTION 5-3

The positive and negative peak voltages of the input signal must remain small enough so the instantaneous input signal voltage produces an *instantaneous operating point* that is always within the linear (straight-line), sloped part of the curve.

PROBLEM 5-4

What conditions can produce nonlinearity in an FET circuit?

SOLUTION 5-4

Improper bias or excessive drive (input signal amplitude), or both, can produce nonlinearity. Under such conditions, as shown in Fig. 5-5, the instantaneous operating point would go either into the region marked *Y* (at left), or into the curved part of the graph (at right), or both, during part of the cycle. This would cause distortion in the output audio.

The MOSFET

The acronym *MOSFET* (pronounced "*moss*-fet") stands for *metal-oxide-semiconductor field effect transistor*. Like JFETs, MOSFETs can be either N-channel or P-channel. A simplified cross-sectional drawing of an N-channel MOSFET, along with the schematic symbol, is shown in Fig. 5-7. The P-channel device is shown in the drawings of Fig. 5-8.

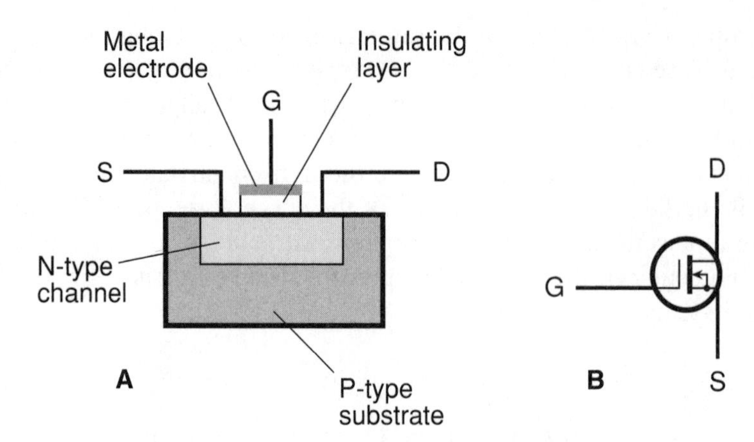

Figure 5-7 At A, pictorial diagram of an N-channel MOSFET. At B, the schematic symbol. Electrodes are S = source, G = gate, and D = drain.

EXTREMELY HIGH INPUT IMPEDANCE

When the MOSFET was first developed, it was called an *insulated-gate FET* or *IGFET*. That's still a good name for it. The gate electrode is insulated from the channel by a thin layer of *dielectric* material. As a result, the input impedance is even higher than that of a JFET. The gate-to-source resistance of a typical MOSFET is, as a matter of fact, comparable to that of a low-loss capacitor! This means that a MOSFET draws essentially no current, and therefore no power,

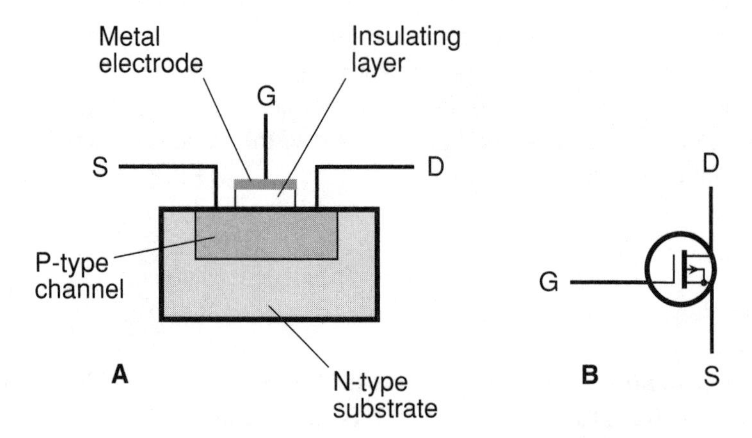

Figure 5-8 At A, pictorial diagram of a P-channel MOSFET. At B, the schematic symbol. Electrodes are S = source, G = gate, and D = drain.

from the signal source. This makes the device even better, in theory, than the JFET for use in low-level and weak-signal amplifiers. But MOS devices are electrically fragile.

BEWARE OF "STATIC"

The trouble with MOSFETs is that they can be easily damaged by *electrostatic discharge*, also informally called "static." When building or servicing circuits containing MOS devices, technicians must use special equipment to ensure that their hands don't carry electrical charges that might ruin the components. If a discharge occurs through the dielectric of a MOS device, the component is permanently destroyed. Humid weather does not offer total protection against the hazard. It doesn't even take a noticeable spark to ruin a MOS component.

FLEXIBILITY IN BIASING

In electronic circuits, an N-channel JFET can sometimes be replaced directly with an N-channel MOSFET, and P-channel devices can be similarly interchanged. But the characteristic curves for MOSFETs are not the same as those for JFETs. The main difference is that the S-G junction in a MOSFET is not a P-N junction. Therefore, forward breakover cannot occur, no matter what the bias. A gate bias voltage, E_G, more positive than +0.6 V can be applied to an N-channel MOSFET, or an E_G more negative than –0.6 V to a P-channel device, without a current "leak" taking place. The other undesirable effects of such bias—nonlinearity and low efficiency—can still occur, however. Figure 5-9 shows a family of characteristic curves for a hypothetical N-channel MOSFET.

DEPLETION MODE VERSUS ENHANCEMENT MODE

Normally the channel in a JFET is open under conditions of no gate bias. As the depletion region gets wider, choking off the channel, the charge carriers are forced to pass through a narrower path. This is known as the *depletion mode* of operation for a field effect transistor. A MOSFET can also be made to work in the depletion mode. The drawings and schematic symbols of Figs. 5-7 and 5-8 show depletion-mode MOSFETs.

Metal-oxide semiconductor technology also allows an entirely different means of operation known as the *enhancement mode*. This type of MOSFET normally

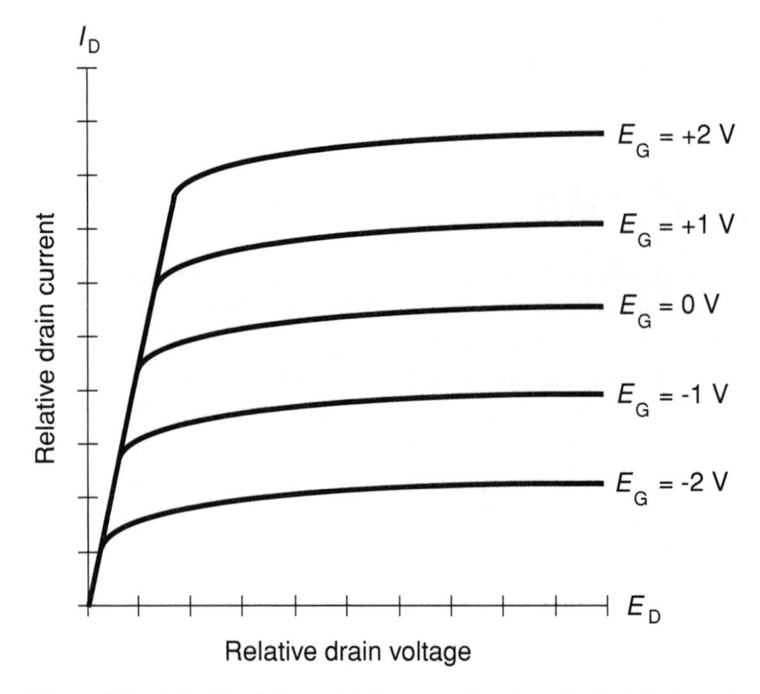

Figure 5-9 A family of characteristic curves for a hypothetical N-channel MOSFET.

has a pinched-off channel when there is no gate bias. It is necessary to apply a bias voltage, E_G, to the gate so that a channel will form. If $E_G = 0$ in an enhancement-mode MOSFET, then $I_D = 0$ when there is no signal input.

The schematic symbols for N-channel and P-channel enhancement mode devices are shown in Fig. 5-10. The vertical line to which the source and drain leads run is broken. This is how you can recognize an enhancement mode device in circuit diagrams.

PROBLEM 5-5

In what types of audio circuit might a MOSFET be used to greatest advantage?

SOLUTION 5-5

Because of their extremely high input impedance, MOSFETs are a good choice for use in preamplifiers for microphones, compact disc (CD) players, and specialized devices such as ultrasonic sensors or SONAR equipment.

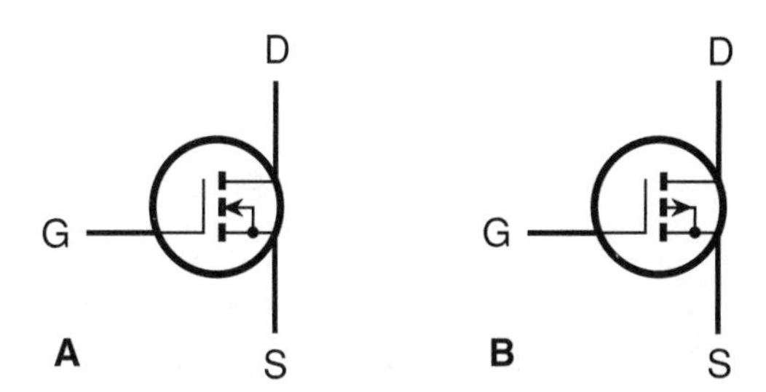

Figure 5-10 Schematic symbols for enhancement-mode MOSFETs. At A, the N-channel device; at B, the P-channel device.

Common-Source Circuit

There are three different circuit hookups for FETs, just as there are for bipolar transistors. These three arrangements place the source, the gate, or the drain at signal ground. In the *common-source circuit*, the source is placed at signal ground. The audio input is applied to the gate. The general configuration is shown in Fig. 5-11. An N-channel JFET is used here, but the device could be an N-channel, depletion-mode MOSFET and the circuit diagram would be the same. For an N-channel, enhancement mode device, an extra resistor would be necessary, running from the gate to the positive power supply terminal. For P-channel devices, the circuit would be the same as that shown here, but the power supply would provide a negative, rather than a positive, DC voltage to the drain.

Capacitor C_1 and resistor R_1 place the source at signal ground while elevating the source above ground for DC. The AC signal enters through C_2. Resistor R_2 determines the impedance "seen" by the preceding stage or input device, and provides bias for the gate. The AC signal passes out of the circuit through C_3. Resistor R_3 keeps the output signal from being shorted out through the power supply.

The common-source arrangement provides the greatest realizable gain of the three FET circuit configurations. For sine-wave signals, the output is 180° out of phase with the input. In general, the output wave is inverted with respect to the input wave.

Common-Gate Circuit

In the *common-gate circuit* (Fig. 5-12), the gate is placed at signal ground. The input is applied to the source. The illustration shows an N-channel JFET. For other types

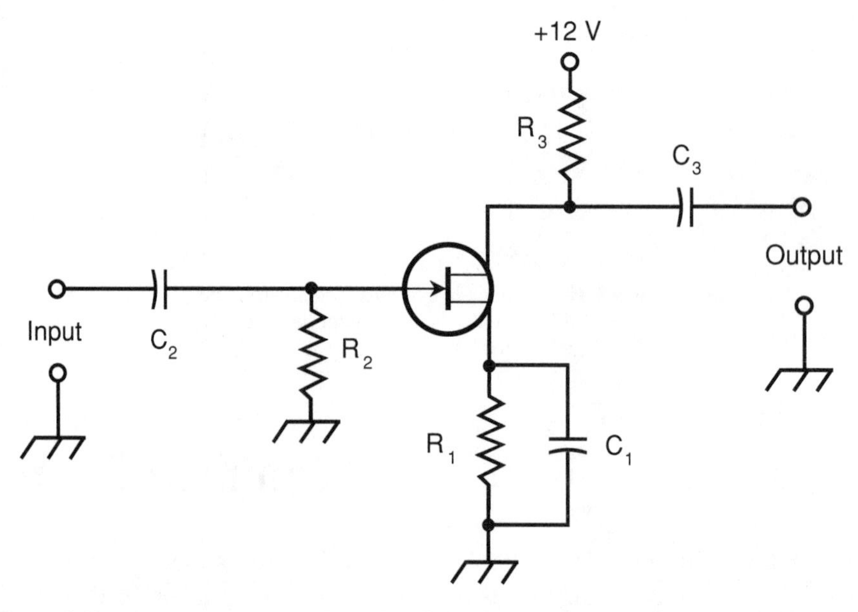

Figure 5-11 Common-source configuration. This diagram shows an N-channel JFET circuit.

of FETs, the same considerations apply as described above for the common-source circuit. Enhancement-mode devices would require a resistor between the gate and the positive supply terminal (or the negative terminal if the MOSFET is P-channel).

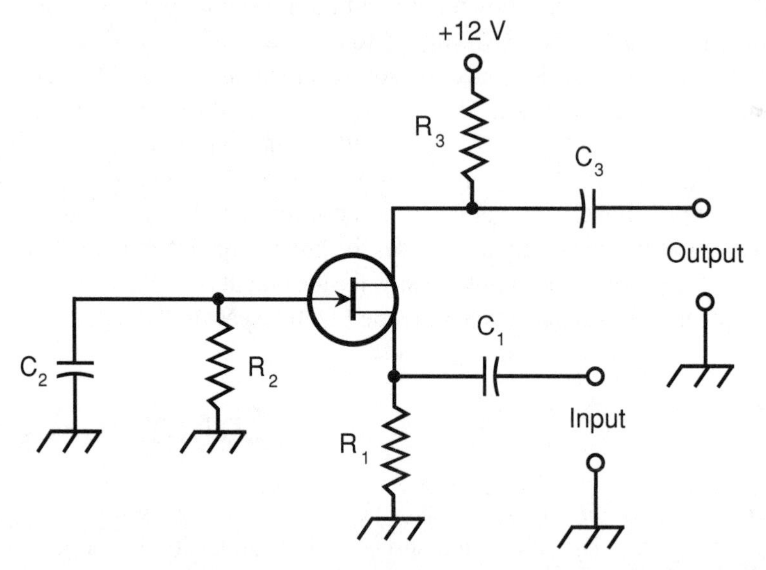

Figure 5-12 Common-gate configuration. This diagram shows an N-channel JFET circuit.

The DC bias for the common-gate circuit is basically the same as that for the common-source arrangement. But the signal follows a different path. The AC input signal enters through C_1. Resistor R_1 keeps the input from being shorted to ground. Gate bias is provided by R_1 and R_2. Capacitor C_2 places the gate at signal ground. In some common-gate circuits, the gate is directly grounded, and R_2 and C_2 are not necessary. The output signal leaves the circuit through C_3. Resistor R_3 keeps the output signal from being shorted through the power supply.

The common-gate arrangement produces less gain than its common-source counterpart. But a common-gate amplifier is not likely to break into unwanted oscillation, making it a good choice for audio power-amplifier circuits. The output is in phase with the input.

Common-Drain Circuit

A *common-drain circuit* is shown in Fig. 5-13. In this circuit, the collector is at signal ground. It is sometimes called a *source follower*. The FET is biased in the same way as for the common-source and common-gate circuits. In the illustration, an N-channel JFET is shown, but any other kind of FET could be used, reversing the polarity for P-channel devices. An enhancement-mode MOSFET would need

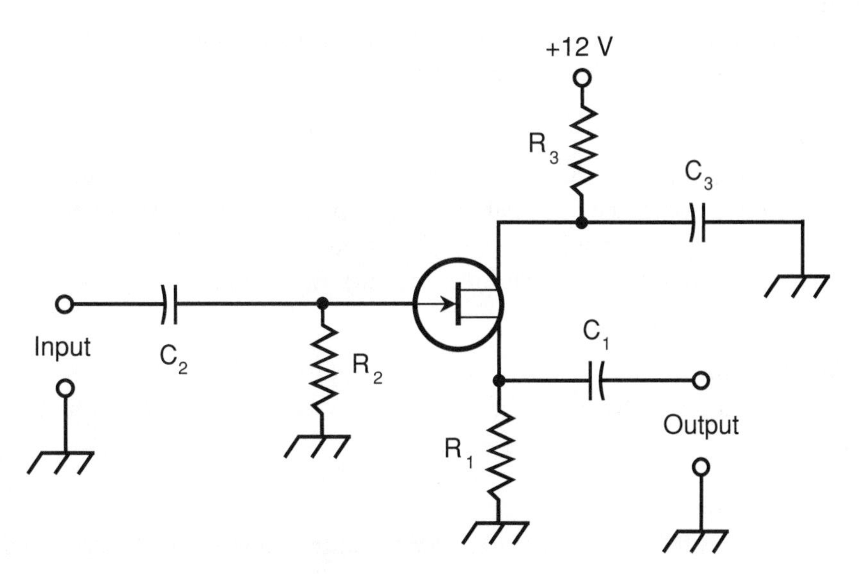

Figure 5-13 Common-drain configuration, also known as a source follower. This diagram shows an N-channel JFET circuit.

a resistor between the gate and the positive supply terminal (or the negative terminal if it is a P-channel device).

The input signal passes through C_2 to the gate. Resistors R_1 and R_2 provide gate bias. Resistor R_3 limits the current. Capacitor C_3 keeps the drain at signal ground. Fluctuating DC (the channel current) flows through R_1 as a result of the input signal; this causes a fluctuating DC voltage to appear across R_1. The output is taken from the source, and its AC component passes through C_1.

The output of the common-drain circuit is in phase with the input. This scheme is the FET analog of the bipolar common-collector arrangement. The output impedance is rather low, making this circuit a good choice for impedance matching over a wide range of frequencies.

PROBLEM 5-6

In the common-source circuit of Fig. 5-11, what would happen if capacitor C_1 were to short out?

SOLUTION 5-6

The circuit would still work, although the gain might be slightly reduced, and in some cases distortion would occur because the shorted-out capacitor would place the source directly at DC ground, altering the bias normally provided by resistor R_1.

PROBLEM 5-7

In the common-source circuit of Fig. 5-11, what would happen if capacitor C_1 were to open up and provide no capacitance at all?

SOLUTION 5-7

The circuit would still work, although the gain would be slightly reduced. The source would no longer be at signal ground. Some common-source amplifiers are designed that way deliberately (without the capacitor C_1), because the diminished gain reduces the risk of the circuit breaking into oscillation.

PROBLEM 5-8

In the common-drain circuit of Fig. 5-13, what would happen if resistor R_2 were to short out?

SOLUTION 5-8

This would short the input signal to ground, and the output would therefore drop to zero.

Quiz

Refer to the text in this chapter if necessary. A good score is at least 8 correct. Answers are in the back of the book.

1. The current through the channel of a JFET is *directly* affected by all of the following, except the
   (a) drain voltage.
   (b) transconductance.
   (c) gate voltage.
   (d) gate bias.

2. In an N-channel JFET, pinchoff occurs when the gate bias voltage is
   (a) sufficiently positive.
   (b) zero.
   (c) sufficiently negative.
   (d) Forget it! Pinchoff never occurs in an N-channel JFET.

3. In a P-channel JFET,
   (a) the drain is forward-biased.
   (b) the source-gate junction is forward-biased.
   (c) the drain is placed at a negative DC voltage relative to the source.
   (d) the gate must be at DC ground.

4. The gate of a MOSFET exhibits a
   (a) forward bias.
   (b) high impedance.
   (c) low reverse resistance.
   (d) low avalanche voltage.

5. When a JFET is biased well beyond pinchoff,
   (a) the value of dI_D/dE_G is positive with a weak-signal input.
   (b) the value of dI_D/dE_G might vary considerably with a weak-signal input.
   (c) the value of dI_D/dE_G is negative with a weak-signal input.
   (d) the value of dI_D/dE_G is zero with a weak-signal input.

6. Characteristic curves for JFETs generally show
   (a) drain voltage as a function of source current.
   (b) drain current as a function of gate current.
   (c) drain current as a function of drain voltage.
   (d) drain voltage as a function of gate current.
7. A significant difference between MOSFETs and JFETs is the fact that
   (a) MOSFETs can handle a wider range of gate bias voltages.
   (b) MOSFETs can deliver greater output power.
   (c) MOSFETs are more rugged.
   (d) MOSFETs last longer.
8. When an enhancement-mode MOSFET is at zero bias,
   (a) the drain current is high with no signal.
   (b) the drain current fluctuates with no signal.
   (c) the drain current is low with no signal.
   (d) the drain current is zero with no signal.
9. In a source follower, which of the electrodes receives the input signal?
   (a) Any of them; it doesn't matter.
   (b) the source.
   (c) the gate.
   (d) the drain.
10. Which of the following circuits can produce the greatest signal gain (amplification factor)?
    (a) The common-source circuit.
    (b) The common-gate circuit.
    (c) The common-drain circuit.
    (d) All of the above circuits can amplify to the same extent.

CHAPTER 6

Electron Tubes

Electron tubes, also called *tubes* or *valves* (in England), are used in some electronic equipment. In a tube, the charge carriers are *free electrons* that travel through space between electrodes inside the device. This makes tubes fundamentally different from semiconductor devices, in which charge carriers move among atoms in a solid medium.

Tube Forms

There are two basic types of electron tube: the *vacuum tube* and the *gas-filled tube*. As their names imply, vacuum tubes have the gases evacuated. Gas-filled tubes contain elemental vapor at low pressure.

VACUUM TUBE

Vacuum tubes accelerate electrons to high speeds, resulting in large currents. In some vacuum tubes, this current can be focused into beams and guided in particular directions. The intensity and/or beam direction can be changed with extreme rapidity, producing effects such as rectification, detection, oscillation, amplification, signal mixing, waveform displays, spectral displays, and video imaging.

GAS-FILLED TUBE

Gas-filled tubes exhibit constant voltage drop (the voltage across them is constant) regardless of the current. This makes them useful as voltage regulators for high-voltage, high-current power supplies. Gas-filled tubes can withstand conditions that would destroy semiconductor devices. Some specialized gas-filled tubes emit infrared (IR), visible light, and/or ultraviolet (UV). This property can be put to use for decorative lighting. A small *neon bulb* can be employed to construct an *audio relaxation oscillator* (Fig. 6-1).

Electrodes in a Tube

In a tube, the electron-emitting electrode is known as the *cathode*. The cathode is usually heated by means of a wire *filament*, similar to the glowing element in an incandescent bulb. The heat from the filament drives electrons from the cathode. The cathode of a tube is analogous to the source of an FET, or to the emitter of a bipolar transistor. The electron-collecting electrode is called the *anode* or *plate*. The plate is the tube counterpart of the drain of an FET or the collector of a bipo-

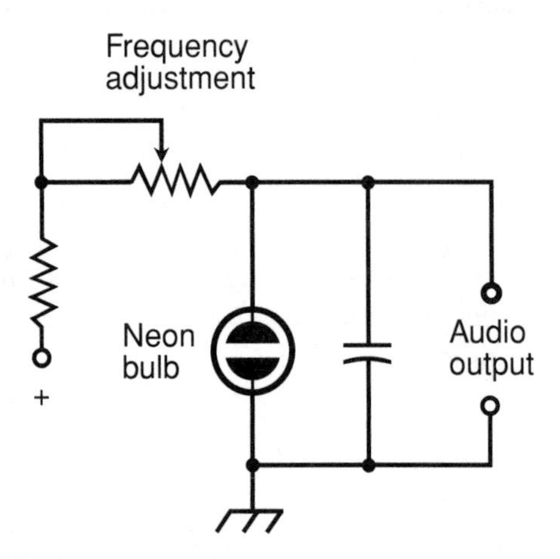

Figure 6-1 A neon bulb oscillator, also known as a relaxation oscillator.

lar transistor. In most tubes, intervening *grids* control the flow of electrons from the cathode to the plate. The grids are the counterparts of the gate of an FET or the base of a bipolar transistor.

DIRECTLY HEATED CATHODE

In some tubes, the filament also serves "double duty" as the cathode. This type of electrode is called a *directly heated cathode*. The negative power supply voltage is applied directly to the filament. The filament voltage for most tubes is 6 V or 12 V DC. It is important that DC be used to heat the filament in this type of tube, because AC will *modulate* the output. In an audio amplifier, this would cause objectionable hum. The schematic symbol for a diode (two-element) tube with a directly heated cathode is shown in Fig. 6-2A.

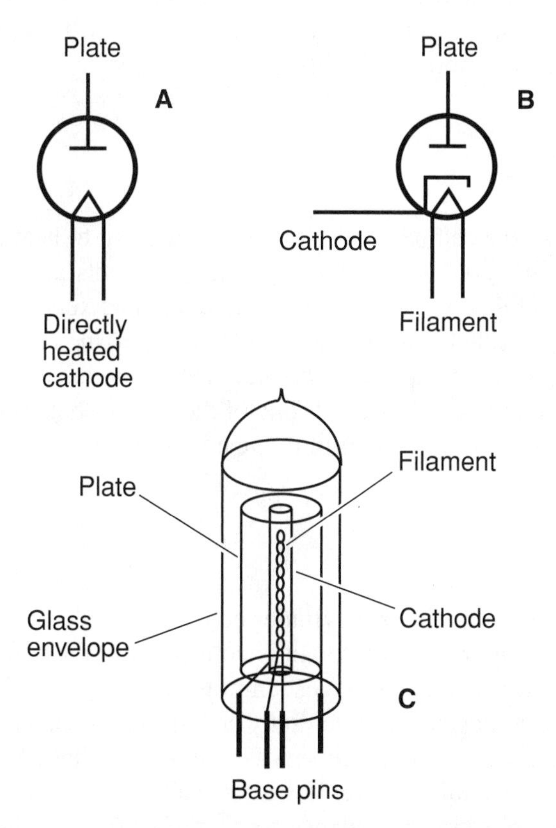

Figure 6-2 At A, schematic symbol for diode tube with directly heated cathode. At B, symbol for diode tube with indirectly heated cathode. At C, simplified rendition of the construction of a diode tube.

INDIRECTLY HEATED CATHODE

In many types of tubes, the filament is enclosed within a cylindrical cathode, and the cathode gets hot from the IR radiated by the filament. This is an *indirectly heated cathode*. The filament normally receives 6 V or 12 V AC or DC. In an indirectly heated cathode arrangement, AC does not cause modulation problems, as it can with a directly heated cathode tube. The schematic symbol for a diode tube with an indirectly heated cathode is shown in Fig. 6-2B.

Because the electron emission in a tube depends on the filament or *heater*, vacuum tubes need a certain amount of time to warm up before they can operate properly from a cold start. This time can vary from a few seconds (for a small tube with a directly heated cathode) to a couple of minutes (for massive power-amplifier tubes with indirectly heated cathodes). The warm-up time for a large tube is about the same as the boot-up time for a personal computer. The warm-up time can be eliminated by keeping the filaments powered-up all the time, even when the circuits that use the tubes are switched off. However, this consumes energy and can shorten tube life.

COLD CATHODE

In a gas-filled tube, the cathode does not have a filament to heat it. Such an electrode is called a *cold cathode*. Various chemical elements are used in gas-filled tubes. In fluorescent devices, elemental neon, argon, and xenon gas are common. In old-fashioned gas-filled voltage-regulator (VR) tubes, mercury vapor was used. A special heating element vaporized the mercury. These tubes are not often seen nowadays, because mercury is toxic. If one of these tubes breaks or is improperly discarded, the mercury can be harmful to the environment.

PLATE

The plate, or anode, of a tube is a cylinder concentric with the cathode and filament, as shown in Fig. 6-2C. The plate is connected to the positive DC supply voltage. Tubes operate at plate voltages ranging from about 50 V to several thousand volts. These voltages are potentially lethal. Technicians unfamiliar with vacuum tubes should not attempt to service equipment that contains them. The output of a tube type amplifier circuit is almost always taken from the plate circuit. The plate exhibits high impedance for signal output, similar to the drain of an FET.

DIODE TUBE

Even before the year 1900, scientists knew that electrons could carry electric current through a vacuum. They also knew that hot electrodes emit electrons more easily than cool ones. These phenomena were put to use in simple diode tubes, for the purpose of *rectification* (converting AC to DC). Diode tubes have two electrodes: a cathode and a plate. Current can flow easily from the cathode to the plate, but not vice-versa. Diode tubes are rarely used nowadays, although they can still be found in some power supplies that are required to deliver several thousand volts for long periods at a 100% *duty cycle* (that is, continuous operation).

CONTROL GRID

The flow of current in a vacuum tube can be controlled by means of an electrode between the cathode and the plate. This electrode, the *control grid* (or simply the *grid*) is a wire mesh or screen that lets electrons pass through. The grid impedes the flow of electrons if it is provided with a negative voltage relative to the cathode. This effect is much the same as the constriction of the channel in an N-channel JFET when a negative voltage is applied to the gate. The greater the negative *grid bias*, the more the grid obstructs the flow of electrons through the tube.

In most cases, the control grid of a vacuum tube is supplied with a negative voltage with respect to the cathode. The negative grid bias is a fraction of the positive plate voltage. However, some tubes are designed to operate best at zero bias, where the grid is at the same DC potential as the cathode.

INTERELECTRODE CAPACITANCE

In a vacuum tube, the internal electrodes exhibit *interelectrode capacitance* that is the primary limiting factor on the frequency range in which the device can produce gain. The interelectrode capacitance in a typical tube is a few picofarads. Vacuum tubes intended for use as audio power amplifiers are designed to minimize this capacitance.

TRIODE TUBE

A tube with one grid is called a *triode* because it has three electrodes. The schematic symbol for a triode tube is shown at Fig. 6-3A. In this case the cathode

is indirectly heated, and the filament is not shown. (This omission is standard in schematics showing tubes with indirectly heated cathodes.) When the cathode is directly heated, the filament symbol serves as the cathode symbol. The control grid is usually biased with a negative DC voltage ranging from near zero to as much as half the positive DC plate voltage.

TETRODE TUBE

A second grid can be added between the control grid and the plate. This is a spiral of wire or a coarse screen, and is called the *screen grid* or *screen*.

The screen, when supplied with a positive DC voltage of 25 to 35 percent of the plate voltage, reduces the interelectrode capacitance between the control grid and plate, minimizing the tendency of a tube type audio power amplifier to break into *oscillation*. When it happens, the tube amplifies stray output signals and produces "squealing," "howling," or radio signals at unpredictable frequencies. Its ability to amplify the desired audio is degraded.

The screen grid can also serve as a second control grid, allowing two signals to be injected into a tube. A tube with two grids has four elements, and is known as a *tetrode*. Its schematic symbol is shown at B in Fig. 6-3.

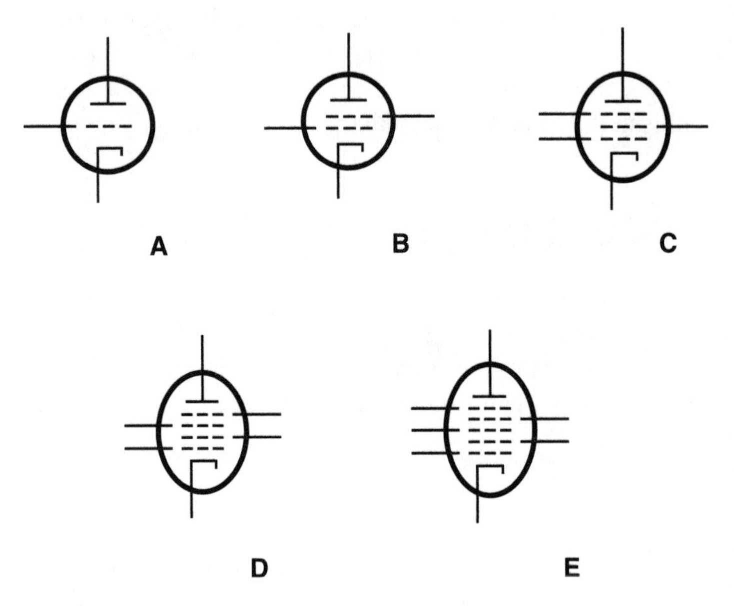

Figure 6-3 Schematic symbols for vacuum tubes with grids: triode (A), tetrode (B), pentode (C), hexode (D), and heptode (E).

PENTODE TUBE

The electrons in a tetrode can bombard the plate with such force that some of them bounce back, or knock other electrons from the plate. This so-called *secondary emission* can hinder tube performance and, at high power levels, cause the *screen current* to become so high that the electrode is destroyed. This problem can be eliminated by placing another grid, called the *suppressor grid* or *suppressor*, between the screen and the plate. The suppressor is typically provided with the same DC voltage as the cathode.

The suppressor repels *secondary electrons* emanating from the plate, preventing most of them from reaching the screen. These secondary electrons are literally "knocked loose" from the plate as the fast-moving electrons from the cathode strike it. The suppressor also reduces the interelectrode capacitance between the control grid and the plate more than a screen grid by itself. A tube with three grids is known as a *pentode*.

The pentode tube design offers greater gain and stability than can be achieved with a tetrode or triode. The schematic symbol for a pentode is shown at C in Fig. 6-3.

HEXODE AND HEPTODE TUBES

In older radio and TV receivers, tubes with four or five grids were sometimes used. These tubes had six and seven elements, respectively, and were called the *hexode* and the *heptode*. The usual function of such tubes was *signal mixing*, in which two signals were combined to produce output signals at frequencies equal to the sum of, and the difference between, the input signal frequencies.

The schematic symbol for a hexode is shown at D in Fig. 6-3; the symbol for a heptode is at E. You will not encounter hexodes and heptodes in modern electronics, because solid-state components are now used for signal mixing.

PROBLEM 6-1

What would happen if the control grid in a vacuum tube were supplied with a DC voltage positive with respect to the cathode?

SOLUTION 6-1

In that case, the control grid would draw excessive current. This could result in physical damage to the tube. Its ability to amplify would also be degraded. It might not amplify at all.

PROBLEM 6-2

If the suppressor in a pentode is at the same voltage as the cathode, won't that grid block all the electrons so none of them can reach the plate?

◆

SOLUTION 6-2

No. The electrons have sufficient speed, and therefore sufficient momentum, to pass through the suppressor on their way from the cathode to the plate. However, the secondary electrons, which are emitted by the plate, move much more slowly. They don't have enough momentum to pass through the suppressor and get back to the screen grid.

PROBLEM 6-3

Suppose a high DC voltage, negative with respect to the cathode, is applied to the suppressor grid in a pentode tube. Can this stop electrons from reaching the plate from the cathode?

SOLUTION 6-3

Yes, if the negative voltage is high enough. The technique can be used to vary the gain of an amplifier. It can also be used to electronically switch the amplifier on and off at a high rate of speed.

Circuit Configurations

The most common audio application of vacuum tubes these days is in power amplifiers. In vintage systems they are common, and a few high-fidelity audio systems still employ them. In recent years, tubes have gained favor with a few popular music bands. Some musicians insist that "tube amps" provide richer sound than amplifiers using power transistors. There are two basic vacuum-tube amplifier circuit arrangements: the *grounded-cathode* configuration and the *grounded-grid* configuration.

GROUNDED-CATHODE CIRCUIT

Figure 6-4 is a simplified schematic diagram of a grounded-cathode circuit using a triode tube. This circuit is the basis for many tube type audio power amplifiers. The input impedance is moderate, and the output impedance is high. In audio power amplifiers, impedance matching between the *driving stage* (the circuit whose output is connected to the amplifier input) and the amplifier, and between the amplifier and the *load* (whatever is driven by the amplifier, such as the speakers), is usually done by means of transformers.

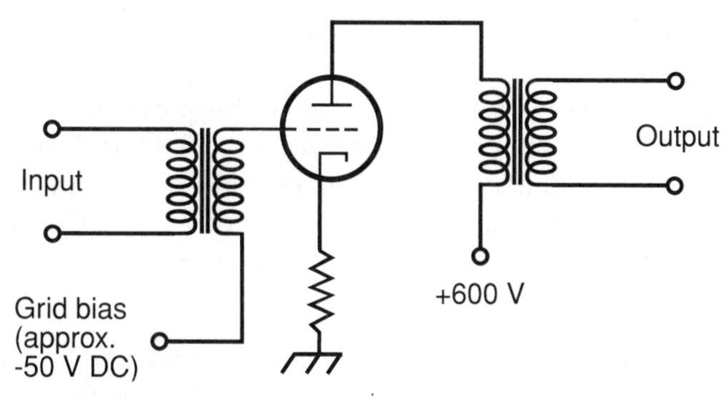

Figure 6-4 Simplified schematic diagram of a grounded-cathode audio power amplifier using a triode tube.

GROUNDED-GRID CIRCUIT

Figure 6-5 shows a basic grounded-grid amplifier circuit. The input impedance is low, and the output impedance is high. The input and output impedances are matched by the same means as with the grounded-cathode arrangement. A grounded-grid amplifier requires more driving power than the grounded-cathode scheme to obtain the same power output. A grounded-cathode amplifier might produce 1 kW of audio output for 10 W input, but a grounded-grid amplifier needs 50 W to 100 W of drive to produce 1 kW of audio output. There is a reward for this increased drive requirement: a grounded-grid amplifier is less likely to break into unwanted oscillation than a grounded-cathode amplifier.

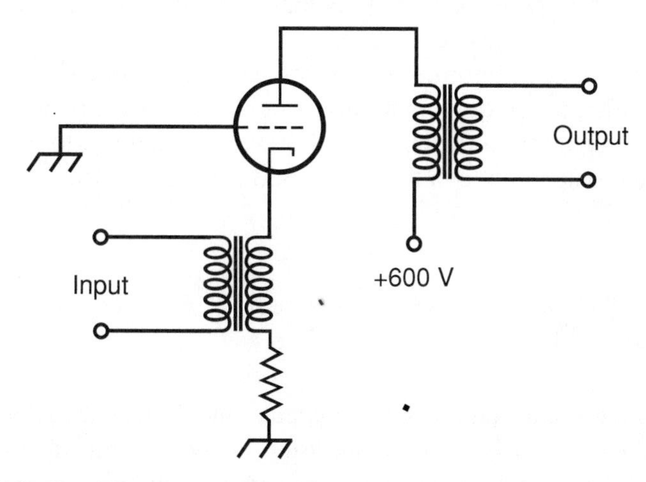

Figure 6-5 Simplified schematic diagram of a grounded-grid audio power amplifier using a triode tube designed to operate at zero bias (grid at same DC potential as cathode).

PLATE VOLTAGE

The plate voltages (+600 V DC) in the circuits of Figs. 6-4 and 6-5 are given as examples. The amplifiers shown could produce 75 W to 150 W of audio output provided they receive sufficient drive and are properly biased. An amplifier rated at 1 kW audio output would require a plate voltage of +2 kV DC to +5 kV DC.

PROBLEM 6-4

What is the purpose of the resistor in series with the cathode in Fig. 6-4? What is the purpose of the resistor in series with the cathode transformer winding in Fig. 6-5?

SOLUTION 6-4

In both instances, the resistor limits the current that can flow through the tube. It is typically a small-value resistor in terms of its ohmic value. It prevents damage to the tube in the event of loss of grid bias (Fig. 6-4) or excessive driving power (Figs. 6-4 and 6-5). This resistor also reduces the risk that the grounded cathode circuit of Fig. 6-4 will break into oscillation.

PROBLEM 6-5

What would happen if the grid in the circuit of Fig. 6-4 were suddenly short-circuited to the cathode inside the tube?

SOLUTION 6-5

This would represent permanent physical damage to the vacuum tube. The negative grid bias voltage would cause a large current to flow directly through the grid, the cathode, the input transformer secondary, and the cathode resistor. In a well-designed amplifier, the power supply fuse would blow or the breaker would trip, removing all voltages from the circuit and preventing damage to the transformer or the cathode resistor, although the tube would have to be replaced.

Cathode-Ray Tubes

Many TV receivers, and some desktop computer monitors, use *cathode-ray tubes* (CRTs). So do older oscilloscopes, spectrum analyzers, and radar sets. You'll come across these if you use test equipment more than a few years old in your audio projects!

ELECTRON BEAM

In a CRT, a specialized cathode called an *electron gun* emits an electron beam that is focused and accelerated as it passes through positively charged *anodes*. The beam then strikes a glass screen whose inner surface is coated with *phosphor*. The phosphor glows visibly, as seen from the face of the CRT, because of the effect of the high-speed electrons striking it.

The beam *scanning* pattern is controlled by magnetic or electrostatic fields. One field causes the beam to scan rapidly across the screen in a horizontal direction. Another field moves the beam vertically. When complex waveforms are applied to the electrodes that produce the deflection of the electron beam, a display pattern results. This pattern can be the graph of an audio wave, a fixed image, an animated image, a computer text display, or any other type of visible image.

ELECTROMAGNETIC CRT

Figure 6-6 is a simplified cross-sectional drawing of an *electromagnetic CRT*. There are two sets of *deflection coils*, one for the horizontal plane and the other for the vertical plane. (To keep the illustration reasonably clear, only one set of coils is shown.) The greater the current in the coils, the greater the intensity of the magnetic fields they produce, and the more the electron beam is deflected. The electron beam is bent at right angles to the magnetic lines of flux.

In a *cathode-ray oscilloscope*, the horizontal deflection coils receive a sawtooth waveform. This causes the beam to scan, or *sweep*, at a precise, adjustable speed across the screen from left to right as viewed from the front. After each timed left-to-right sweep, the beam returns, almost instantly, to the left side of the screen for the next sweep. The vertical deflection coils receive the waveform to be analyzed. This waveform makes the electron beam move up and down. The combination of vertical and horizontal beam motion produces a display of the input waveform as a function of time.

ELECTROSTATIC CRT

In an *electrostatic CRT*, charged metal plates, rather than current-carrying coils, are used to deflect the electron beam. When voltages appear on these *deflection plates*, the beam is bent in the direction of the electric lines of flux. The greater the voltage applied to a deflection plate, the stronger the electric field, and the greater the extent to which the beam is deflected.

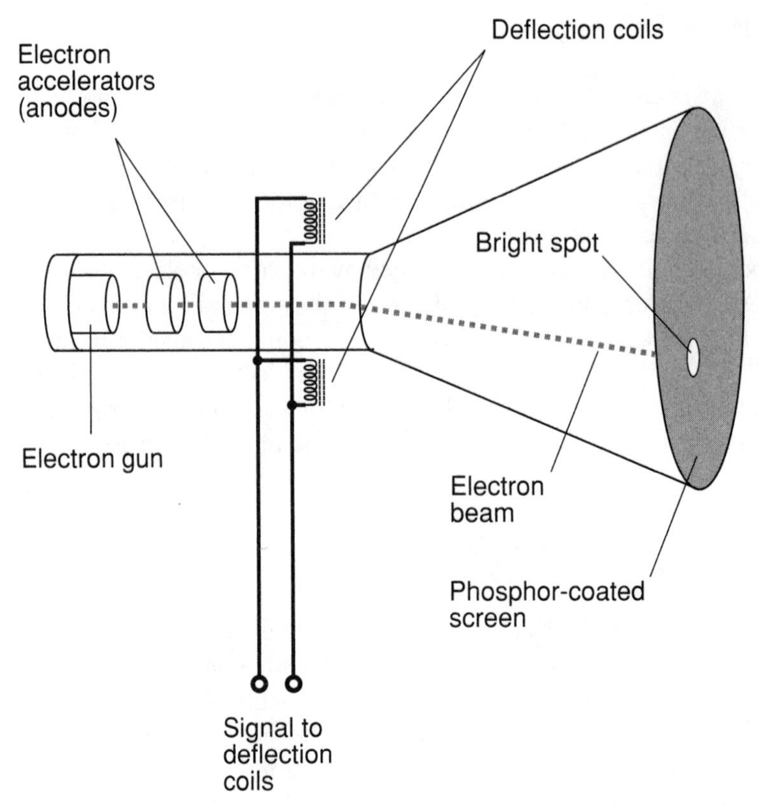

Figure 6-6 Simplified cross-sectional rendition of an electromagnetic CRT, showing one pair of deflection coils.

The principal advantage of an electrostatic CRT is the fact that it generates a far less intense magnetic field than an electromagnetic CRT. This so-called *extremely-low-frequency* (ELF) energy may have adverse effects on people who use electromagnetic CRT-equipped devices, such as desktop computers, for extended periods of time. In recent years, with the evolution of *liquid-crystal displays* (LCDs) and *plasma displays* as alternatives to the CRT type of display, ELF has become a less significant concern.

PROBLEM 6-6

How can an oscilloscope be used to check an audio amplifier for suspected distortion?

SOLUTION 6-6

A *dual-trace oscilloscope* (which has two independent vertical signal inputs) is ideal for this application. The amplifier input signal is applied to one vertical input,

producing a waveform in the top half of the display. The amplifier output signal is applied to the other vertical input, producing a waveform in the bottom half of the display. Variable attenuators are used between the signal sources and the vertical inputs, and are adjusted so the peak-to-peak amplitudes of the two signals are equal. The scope sweep frequency is adjusted to a value such that the waveform details clearly show up. If the amplifier is distortion-free, the two waveforms appear identical on the scope display. If there is any difference in the waveforms, distortion is indicated. This method will not display all types of distortion, but it can show an especially severe form of distortion called *flat topping* (in which the peaks of the waveforms are blunted or flattened). This distortion tends to occur when the amplifier input signal is too strong.

PROBLEM 6-7

Is there another way to check an amplifier for distortion using a scope?

SOLUTION 6-7

Yes. The amplifier input signal can be applied to the scope horizontal input, and the amplifier output signal can be applied to the vertical input. The attenuators (described above) are adjusted so the input and output signals have equal peak-to-peak amplitudes. The signals should be either in phase or 180° out of phase. If the amplifier is distortion-free, the resulting display appears as a straight, sloped line on the screen. Curvature in the trace indicates distortion. This method can indicate any form of signal distortion, and is a more sensitive scheme than the one described in Solution 6-6.

Camera Tubes

Some video cameras use electron tubes that convert visible light into varying electric currents. The two most common types of *camera tube* are the *vidicon* and the *image orthicon*. You should expect to come across these if you intend to create video, *audio-visual*, or *multimedia* productions, so it's a good idea to be familiar with them. There are other forms of video camera devices, but the following are the two most common vacuum tube types.

VIDICON

In the vidicon, a lens focuses the incoming image onto a photoconductive screen. An electron gun generates a beam that sweeps from left to right across the screen as a result of the effects of deflection coils, in a manner similar to the operation of

an electromagnetic CRT. Multiple sweeps occur in parallel horizontal lines from the top of the screen to the bottom, forming a *raster scan* of the screen. The sweep in the vidicon is synchronized with the sweep of any CRT that happens to be displaying the same image.

As the electron beam scans the photoconductive surface, the screen becomes charged. The rate of discharge in a certain region on the screen depends on the intensity of the visible light striking that region. A simplified cutaway view of a vidicon tube is shown in Fig. 6-7.

The main advantage of the vidicon is its small physical size and mass. A vidicon is sensitive, but its response can be sluggish when the level of illumination is low. This causes images to persist for a short while, resulting in poor portrayal of fast-motion scenes.

IMAGE ORTHICON

Another type of camera tube, also quite sensitive but having a quicker response to image changes, is the image orthicon. It is constructed much like the vidicon,

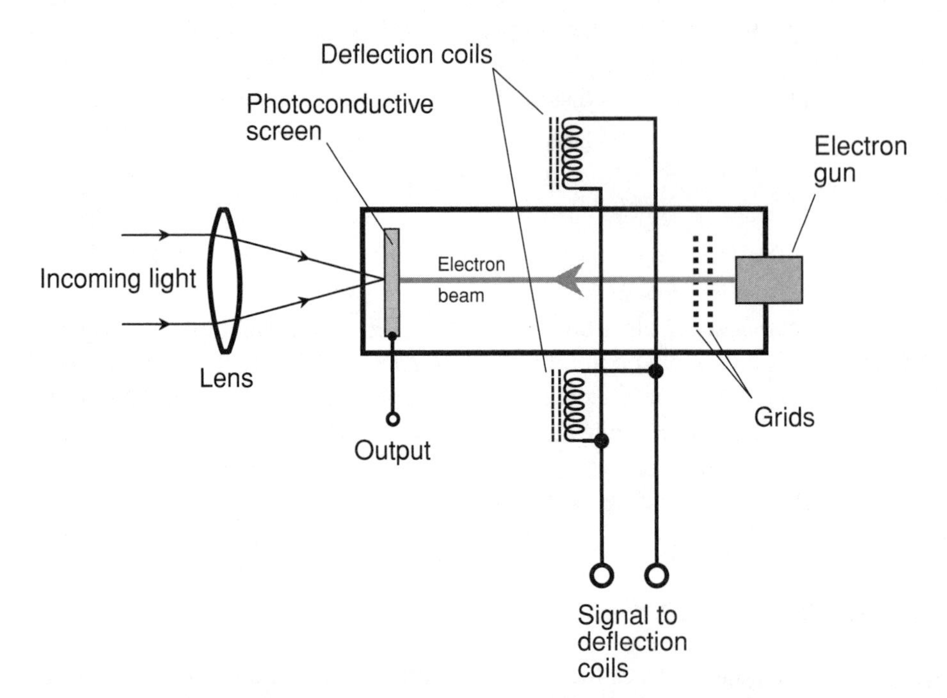

Figure 6-7 Functional diagram of a vidicon, showing one pair of deflection coils.

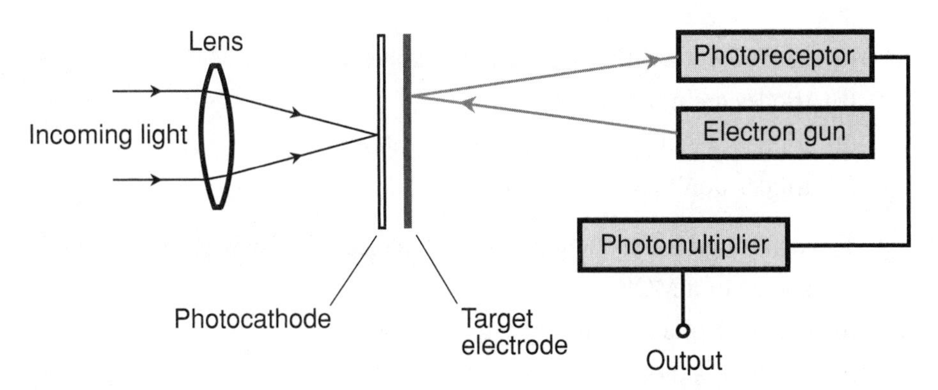

Figure 6-8 Functional diagram of an image orthicon. For simplicity and clarity, the deflection coils are not shown.

except that there is a target electrode behind the *photocathode* (Fig. 6-8). When a single electron from the photocathode strikes the target electrode, multiple secondary electrons are emitted as a result. The image orthicon thus acts as a video signal amplifier, in addition to its function as a camera.

A fine beam of electrons, emitted from the electron gun, scans the target electrode. The secondary electrons cause some of this beam to be reflected back toward the electron gun. Areas of the target electrode with the most secondary-electron emission produce the greatest return beam intensity, and regions with the least emission produce the lowest return beam intensity. The greatest return beam intensity corresponds to the brightest parts of the video image. The return beam is modulated as it scans the target electrode and is picked up by a receptor electrode.

A disadvantage of the image orthicon is that it produces considerable noise in addition to the signal output. But when a fast response is needed and the level of illumination ranges from dim to very bright, the image orthicon is the camera tube of choice.

Quiz

Refer to the text in this chapter if necessary. A good score is at least 8 correct. Answers are in the back of the book.

1. A major difference between a triode tube and an N-channel FET is the fact that
   (a) triodes work with lower signal voltages.
   (b) triodes are more compact.
   (c) triodes need higher power-supply voltages.
   (d) triodes don't need filaments.

2. The control grid of a vacuum tube is the electrical counterpart of the
   (a) source of a MOSFET.
   (b) collector of a bipolar transistor.
   (c) anode of a diode.
   (d) gate of an FET.

3. A screen grid enhances tube operation by
   (a) increasing the gain, helping the circuit to oscillate more easily.
   (b) decreasing the plate voltage required to produce oscillation.
   (c) minimizing the risk that a tube amplifier will break into oscillation.
   (d) pulling excess electrons from the plate.

4. A tube with three grids is called a
   (a) diode.
   (b) triode.
   (c) tetrode.
   (d) pentode.

5. A tube with no grids is called a
   (a) diode.
   (b) triode.
   (c) tetrode.
   (d) pentode.

6. The electron gun in a CRT is another name for its
   (a) cathode.
   (b) anode.
   (c) control grid.
   (d) screen grid.

7. The electron beam in an electrostatic CRT is bent by
   (a) magnetic fields produced by current-carrying coils.
   (b) electric fields produced by charged electrodes.
   (c) a variable voltage on the screen grid.
   (d) visible light striking the electrodes.

8. A vidicon camera tube is noted for its
   (a) poor signal-to-noise ratio.
   (b) large size and heavy weight.
   (c) slow response to image movement in dim light.
   (d) poor sensitivity.

9. The plate in a tetrode tube is normally connected to
   (a) a positive DC power supply voltage.
   (b) a negative DC power supply voltage.
   (c) electrical ground.
   (d) the filament.

10. The screen grid in a tetrode tube is normally connected to
    (a) a positive DC power supply voltage.
    (b) a negative DC power supply voltage.
    (c) electrical ground.
    (d) signal ground.

CHAPTER 7

Audio Characteristics and Components

This chapter is an overview of some fundamental principles governing audio systems. After this chapter, we'll explore how the individual parts of an audio system work.

Properties of Sound

Acoustics is the science of sound waves. Sound consists of molecular vibrations at *audio frequency* (AF), ranging from about 20 Hz to 20 kHz. Young people can hear the full range of AF acoustic waves. As people age, they lose hearing sensitivity at the upper and lower extremes. An elderly person might only hear sounds at frequencies between, say, 40 Hz and 7 kHz.

AUDIO FREQUENCIES

In music, the AF range is divided into three broad, vaguely defined parts, called *bass* (pronounced "base"), *midrange*, and *treble*. The bass frequencies start at 20 Hz and extend to about 200 Hz. Midrange begins at about 200 Hz and extends up to about 3 kHz. Treble consists of the audio frequencies higher than approximately 3 kHz. As the frequency increases, the wavelength of sound in air becomes shorter.

In dry air at sea level and normal barometric pressure, sound travels at about 1100 feet per second (ft/s), or 335 meters per second (m/s). The relationship between the frequency f of a sound wave in hertz, and the wavelength λ_{ft} in feet, is as follows:

$$\lambda_{ft} = 1100/f$$

The relationship between f in hertz and λ_m in meters is given by:

$$\lambda_m = 335/f$$

This formula is also valid for frequencies in kilohertz and wavelengths in millimeters.

An acoustic disturbance at 20 Hz has a wavelength of 55 ft (17 m). A sound of 1000 Hz produces a wave measuring 1.1 ft (34 cm). At 20 kHz, a sound wave in air is only 0.055 ft (17 mm) long. These formulas are accurate enough for most situations involving sound waves in the atmosphere at sea level and at reasonable elevations, but they do not apply in the upper atmosphere, nor do they apply in other materials such as water or metal.

LONGITUDINAL VERSUS TRANSVERSE WAVES

When acoustic waves travel through matter, individual atoms or molecules vibrate to carry the wave energy along. This vibration can be either in line with the direction of wave travel, or at right angles to it. An atom or molecule rarely moves back and forth over a distance of more than a few meters (sometimes less than a millimeter), but the wave overall can propagate for many kilometers.

When an acoustic wave occurs in air or some other gaseous medium, the atoms or molecules vibrate to and fro in line with the direction of wave travel. A wave of this sort is known as a *longitudinal wave*, also called a *compression wave*. In most incompressible or solid materials, the particles move at right angles to the direction of propagation. This is called a *transverse wave*. On the surface of a liquid such as water, the motion of the molecules is both longitudinal and transverse. This can be demonstrated in a lab machine called a *ripple tank*. Suspended particles in

the water describe vertical ellipses in planes parallel to the direction of the wave propagation.

In the applications that concern us, air is always the medium of propagation. The sound waves of interest to hi-fi audio enthusiasts are therefore always of the longitudinal type.

The actual to-and-fro displacement of the atoms or molecules in an acoustic wave is not the same as the wavelength. A weak low-frequency sound wave, for example, will cause a to- and-fro molecular displacement that is much less than the theoretical wavelength. The displacement is, however, related to the loudness of the disturbance. Given an acoustic wave of constant frequency, the to-and-fro displacement of the atoms or molecules increases as the loudness increases.

WAVEFORMS

The loudness (also called *amplitude*, *volume*, or *intensity*) and the frequency (also called the *pitch*) of a sound are not the only variables that acoustic waves can possess. Another important factor is the shape of the wave itself. The simplest acoustic waveform is a sine wave (or *sinusoid*), in which all of the energy is concentrated at a single frequency. Sinusoidal sound waves are rare in nature. A good artificial example is the beat note, or *heterodyne*, produced by a steady carrier in a communications receiver.

In music, most of the notes are complex waveforms, consisting of energy at a specific fundamental frequency and its harmonics. Examples are sawtooth, square, and triangular waves. The shape of the waveform depends on the distribution of energy among the fundamental and the harmonics. A wave with a given fundamental frequency, such as 1 kHz, can have infinitely many different shapes. This gives them a different aspect; they "sound different." For example, a trombone playing middle C sounds much different than a clarinet playing middle C, which in turn sounds different than a violin playing middle C. This aspect of wave, dependent on waveform alone, is called *timbre*. (Sometimes it's called "tone," but technically that is a misnomer.)

PATH EFFECTS

The waveform, as well as the frequency, of an acoustic disturbance affects the manner and extent to which the waves are reflected from solid, semisolid, liquid, or foamy objects. Acoustics engineers must consider this when designing sound systems and concert halls. The goal is to make sure that all the instruments sound realistic everywhere in the room. Theoretical models can help here, but they only go so far. Trial-and-error is necessary. In the end, judgment is made subjectively by diverse listeners.

Suppose you have a sound system set up in your living room, and that, for the particular placement of speakers with respect to your ears, sounds propagate well at 1, 3, and 5 kHz, but poorly at 2, 4, and 6 kHz. This affects the way musical instruments sound. It distorts the sounds from some instruments more than the sounds from others. Unless all sounds, at all frequencies, reach your ears in the same proportions that they come from the speakers, you do not hear the music the way it originally came from the instruments.

Figure 7-1 on page 114 shows a listener, a speaker, and three sound reflectors, also known as *baffles*. The waves X, Y, and Z reflected by the baffles, along with the direct-path wave D, add up to something different at the listener's ears for each frequency of sound. The way they combine also changes as the listener moves around the room. This phenomenon is impossible to prevent. That is why it is so difficult to design an acoustical room, such as a concert auditorium, to propagate sound well at all frequencies for every listener.

PROBLEM 7-1

What is the wavelength in air of a sound wave that has a frequency of 660 Hz? Express the answer in feet and inches, and also in meters.

SOLUTION 7-1

Set $f = 660$ and use the above formulas:

$$\begin{aligned}\lambda_{\text{ft}} &= 1100/f \\ &= 1100/660 \\ &= 1.67 \text{ ft} \\ &= 1 \text{ ft } 8 \text{ in}\end{aligned}$$

$$\begin{aligned}\lambda_{\text{m}} &= 335/f \\ &= 335/660 \\ &= 0.508 \text{ m}\end{aligned}$$

Loudness and Phase

You do not perceive the loudness of sound in direct proportion to the power contained in the disturbance. Instead, your ears and brain sense sound levels according to the *logarithm* of the actual intensity. Another variable is the *phase* with

which waves arrive at your ears. Phase allows you to perceive the direction from which a sound is coming, and it also affects perceived sound volume.

THE DECIBEL IN ACOUSTICS

Sound loudness is usually measured in units called *decibels* (dB). If you change the volume control on a hi-fi set so you can just barely tell the difference in the loudness, the increment is approximately 1 dB. In acoustic applications, decibels express *relative sound power*. If you use the volume control to double the actual sound power coming from a set of speakers, then that is, by definition, a change of +3 dB. Conversely, if you halve the sound power, it is a change of –3 dB. Increases in sound power have positive decibel values, and decreases in sound power are indicated by negative decibel values.

For decibels to have meaning in acoustics, there must be a reference volume level against which all sounds are measured. Have you been told that an electric vacuum cleaner produces 80 dB of sound as heard by the person operating it? This is determined with respect to the *threshold of hearing*, which is the faintest sound that a person with good hearing can detect in a *quiet room* specially designed to have a minimum of background noise.

Suppose P_t is the sound power of an acoustic wave that appears at the threshold of hearing, and P_s is the sound power actually impinging on your eardrums from a specific source such as a vacuum cleaner, musical instrument, or hi-fi set. Then the sound level (let's call it S) in decibels is given by this formula:

$$S = 10 \log (P_s / P_t)$$

The expression "log" refers to the *base-10 logarithm function*. It can be found on any good scientific calculator, including the one provided with most personal computers. The decibel is also used to define relative current, voltage, or power in electronic circuits, particularly amplifiers. You'll learn about that later in this book.

PHASE IN ACOUSTICS

Even if there is only one sound source, acoustic waves reflect from the walls, ceiling, and floor of a room. In the scenario shown by Fig. 7-1, imagine the baffles as two walls and the ceiling in a room. As is the case with baffles, the three sound paths X, Y, and Z have different lengths, so the sound waves reflected from these surfaces do not arrive in the same phase at the listener's ears. The direct path (D), a straight line from the speaker to the listener, is always the shortest path. In this

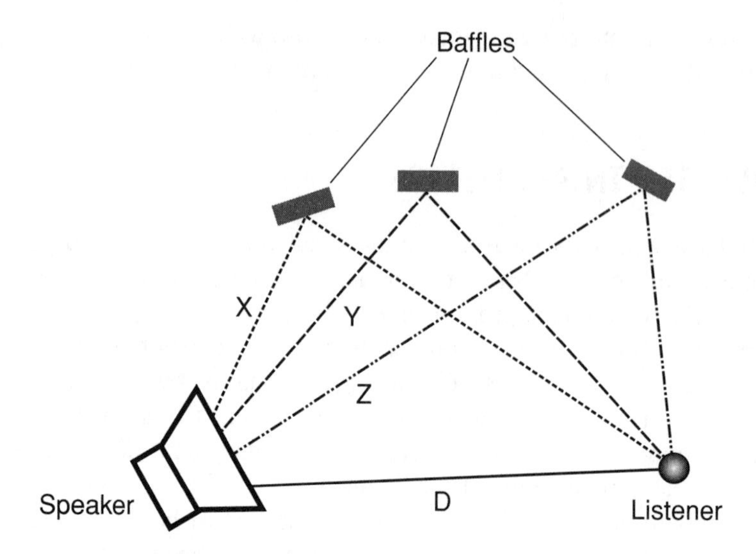

Figure 7-1 Sounds from reflected paths (such as X, Y, and Z) combine with direct-path sound (D) to produce what a listener hears.

situation, there are four different paths by which sound waves can propagate from the speaker to the listener. In some practical scenarios, there are dozens.

Suppose that, at a certain frequency, the acoustic waves for all four paths happen to arrive in exactly the same phase in the listener's ears. Sounds at that frequency are exaggerated in volume. The same phase coincidence can also occur at harmonics of this frequency. This is undesirable because it causes acoustic peaks, called *antinodes*, distorting the original sound. At certain other frequencies, the waves mix in phase opposition. This produces acoustic nulls called *nodes* or *dead zones*. If the listener moves a few feet or even a few inches, the volume at any affected frequency changes, sometimes dramatically. As if this isn't bad enough, a new antinode or node can then present itself at another set of frequencies.

One of the biggest challenges in acoustical design is the avoidance of significant antinodes and nodes. In a home hi-fi system, this can be as simple as minimizing the extent to which sound waves reflect from the ceiling, the walls, the floor, and the furniture. *Acoustical tile* can be used on the ceiling, the walls can be covered with *cork tile*, the floor can be carpeted, and the furniture can be upholstered with cloth. In large auditoriums and music halls, the problem becomes more complex because of the larger sound propagation distances involved, and also because of the fact that sound waves reflect from the bodies of the people in the audience!

PROBLEM 7-2

Imagine two sound waves X and Y having the same frequency, coming from two different speakers equidistant from a listener. Suppose wave X is produced by 10 W of audio power, while wave Y is produced by 40 W. Further suppose that both speakers are identical, so they have the same sound reproduction efficiency. How much louder, in decibels, is the sound produced by wave Y, as compared with the sound produced by wave X?

SOLUTION 7-2

Use the formula for calculating decibels in terms of power. Replace P_t by P_x (for sound wave X, the weaker wave), and replace P_s by P_y (for sound wave Y, the stronger wave). Then the difference (let's call it D_{xy}) in volume between the two waves, in decibels, is:

$$\begin{aligned} D_{xy} &= 10 \log (P_y / P_x) \\ &= 10 \log (40/10) \\ &= 10 \log (4) \\ &= 10 \times 0.602 \\ &= 6.02 \text{ dB} \end{aligned}$$

This can be rounded to 6 dB. That is to say, wave Y is 6 dB louder than wave X.

PROBLEM 7-3

If sound power increases by a factor of 100, how much of a change is that, in decibels? How about a power increase of 1000 times?

SOLUTION 7-3

In these cases, the ratios of output to input power are 100 and 1000, respectively. The base-10 logarithm of 100 is 2, and the base-10 logarithm of 1000 is 3. Multiplying each of these by 10, we see that a 100-fold sound power increase represents a 20 dB increase in loudness, while a 1000-fold sound power increase represents a 30 dB increase in loudness.

Home Audio Systems

A true *audiophile* ("sound lover") assembles a complex system over a period of time, not all at once. In that way, the design ends up best suited to the user's unique needs. Here are some basic considerations that can serve as guidelines when choosing system components.

CONFIGURATIONS

The simplest type of home stereo system is contained in a single box, with an AM/FM radio receiver called a *tuner* along with a *compact disc* (CD) player. The speakers are generally external, but the connecting cables are short. The assets of a so-called *compact hi-fi system* are small size and low cost.

More sophisticated hi-fi systems have separate boxes containing components such as the following:

- An AM tuner
- An FM tuner
- An amplifier or pair of amplifiers
- A CD player
- A computer and its peripherals

The computer is optional, but it facilitates downloading music files or *streaming audio* from the Internet, creating ("burning") CDs, and composing and editing electronic music. A *satellite radio* receiver, a *tape player*, a *turntable*, or other nonstandard peripheral may also be included. The individual hardware units in this type of system, known as a *component hi-fi system*, are interconnected with shielded cables. A component system costs more than a compact system, but the sound quality is better, you get more audio power, you can do more tasks, and you can tailor the system to your preferences.

Some hi-fi manufacturers build all their equipment cabinets to a standard width so they can be mounted one above the other in a *rack*. A so-called *rack-mounted hi-fi system* saves floor space and makes the system look professional. The rack can be mounted on wheels so the whole system, except for the external speakers, can be rolled from place to place.

Figure 7-2 is a block diagram of a typical home stereo hi-fi system. The amplifier *chassis*, on which all the electronic components are mounted, should be grounded to minimize hum and noise, and to minimize susceptibility to interference from external sources. In most systems, the AM antenna is a loopstick built into the cabinet or mounted on the rear panel. The FM antenna can be an indoor type, such as television "rabbit ears," or a directional outdoor antenna equipped with lightning protection hardware.

THE TUNER

A typical tuner can receive signals in the standard AM broadcast band (535 to 1605 kHz) and/or the standard FM broadcast band (88 to 108 MHz). Tuners don't have

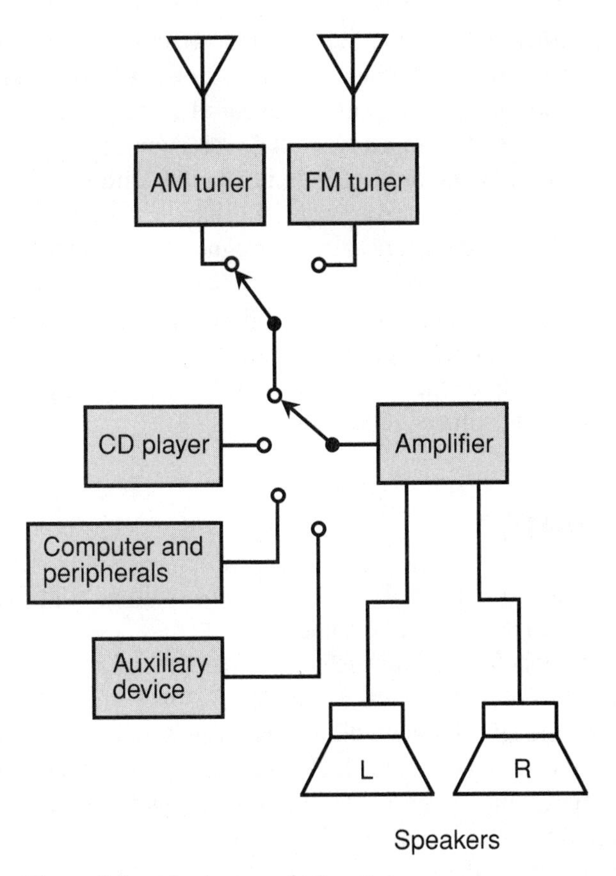

Figure 7-2 A basic stereo hi-fi system.

built-in amplifiers. A tuner can provide enough power to drive a headset, but an amplifier is usually necessary to provide sufficient power to a pair of speakers.

Modern hi-fi tuners employ *frequency synthesizers* and have *digital displays*. Most tuners have several *memory channels*. These are programmable, and allow you to select your favorite stations with a push of a single button, no matter where the stations happen to be in the frequency band. Most tuners also have *seek* and/or *scan* modes that allow the radio to automatically search the band for any station strong enough to be received clearly.

SPEAKERS AND HEADSETS

No amplifier can deliver sound that is better than the speakers will allow. Speakers are rated according to the audio power they can handle. It's a good idea to purchase speakers that can tolerate at least twice the audio output power that the amplifier can deliver. This will ensure that speaker distortion will not occur during loud, low-frequency sound bursts. It will also prevent physical damage to the speakers that might otherwise result from accidentally overdriving them.

Good speakers contain two or three individual units within a single cabinet. The *woofer* reproduces bass. The *midrange speaker* handles medium and, sometimes, treble (high) audio frequencies. A *tweeter* is designed especially for enhanced treble reproduction.

Headsets are rated according to how well they reproduce sound. This is a subjective consideration. Equally expensive headsets can, and often do, exhibit huge differences in the quality of the sound that they put out. Not only that, but people disagree about what constitutes good sound.

BALANCE CONTROL

In hi-fi stereo sound equipment, the *balance control* allows adjustment of the relative volumes of the left and right channels.

In a basic hi-fi system, the balance control consists of a single rotatable knob connected to a pair of *potentiometers* (variable resistors). When the knob is rotated counterclockwise, the left-channel volume increases and the right-channel volume decreases. When the knob is rotated clockwise, the right-channel volume increases and the left-channel volume decreases. In more sophisticated sound systems, the balance is adjusted by means of two independent volume controls, one for the left channel and the other for the right channel.

Proper balance is important in stereo hi-fi. A balance control can compensate for such factors as variations in speaker placement, relative loudness in the channels, and the acoustical characteristics of the room in which the equipment is installed.

TONE CONTROL AND ROLL-OFF

The amplitude vs. frequency characteristics of a hi-fi sound system are adjusted by means of a so-called *tone control*, even though this is technically a misnomer (a better term would be *frequency-response control*). In its simplest form, a tone control consists of a single rotatable knob or linear-motion sliding lever. The counterclockwise, lower, or left-hand settings of this control result in strong bass and weak treble audio output. The clockwise, upper, or right-hand settings result in weak bass and strong treble. When the control is set to mid-position, the audio response of the amplifier is more or less flat; that is, the bass, midrange, and treble are in roughly the same proportions as in the recorded or received signal.

Figure 7-3A shows one way in which a single-potentiometer tone control can be incorporated into the output circuit of an audio amplifier. The amplifier is designed so that its treble output is exaggerated. Potentiometer X attenuates the treble to a variable extent. When the potentiometer is at zero resistance, the capacitor appears directly in parallel with the audio signal, and this causes a *treble roll-off* (decreasing volume with increasing frequency). As the resistance of the potentiometer increases, the treble roll-off becomes less pronounced. At the midpoint of the potentiometer setting, the treble roll-off caused by the *resistance-capacitance* (*RC*) combination cancels out the effect of the exaggerated treble response in the amplifier itself, and the frequency response is flat. When the potentiometer is at its maximum resistance, there is practically no capacitance in parallel with the audio signal. Then the RC combination appears essentially as an open circuit, and the treble response is exaggerated. This type of control also causes a general increase in output volume as the resistance of the potentiometer increases.

A more versatile tone control has two capacitors and two potentiometers, as shown in Fig. 7-3B. The amplifier is designed so that it has a flat amplitude vs. frequency output curve. One *RC* combination is in series, and the other is in parallel.

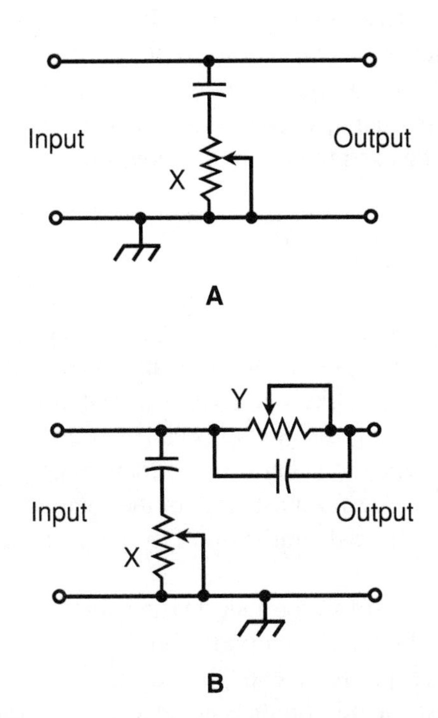

Figure 7-3 Simple methods of obtaining tone control. At A, a single RC combination (X) provides treble attenuation only. At B, one RC combination (X) attenuates the treble, and the other (Y) attenuates the bass.

The series-connected *RC* circuit, labeled X, is connected in parallel with the audio output, and it produces adjustable treble roll-off, exactly as does the series *RC* circuit in illustration A. When Potentiometer X is at zero resistance, the capacitor appears directly in parallel with the audio signal, and this causes a decreasing volume with increasing frequency. As the resistance of the potentiometer increases, the treble roll-off becomes less pronounced. When the potentiometer is at its maximum resistance, there is practically no capacitance in parallel with the audio signal. Then the RC combination appears essentially as an open circuit, and there is no treble roll-off. The parallel *RC* circuit, labeled Y, is in series with the audio path, and it attenuates the bass to a variable extent. When Potentiometer Y is at its maximum resistance, the capacitor across it appears in series with the audio signal, and this causes a *bass roll-off* (decreasing volume with decreasing frequency). As the resistance of the potentiometer decreases, the effect of the capacitor diminishes, and the roll-off becomes less pronounced. When the potentiometer is at zero resistance, there is no capacitance in series with the audio signal because the capacitor is shorted out; there is no bass roll-off. The two potentiometers X and Y can be adjusted separately.

In either of the arrangements shown in Fig. 7-3, the optimum values of the potentiometers and capacitors must be found by experimentation. Of course, separate tone control networks must be installed in both stereo channels. More sophisticated tone controls make use of specialized *integrated circuits* (ICs) called *operational amplifiers* or *op amps*. These can be adjusted to have almost any desired amplitude-vs.-frequency characteristic. In some applications, *digital signal processing* (DSP) can be used to obtain tone control.

AUDIO MIXER

If you simply connect two or more audio sources to the same input of an amplifier, you can't expect good results. Different signal sources, such as a computer, a tuner, and a CD player, are likely to have different effective resistances for AF signals. (The resistance for an AC signal is called *impedance*, and is discussed in Chapter 9.) When connected together, the impedances appear in parallel. This can cause *impedance mismatches* for most or all of the sources, as well as at the amplifier input. The result will be degradation of system efficiency and poor overall performance.

Another problem arises from the fact that the signal amplitudes from various sources almost always differ. A microphone produces minuscule audio-frequency currents, whereas a tuner produces enough to drive a headset. Connecting both of these together will cause the microphone signal to be obliterated by the signal from the tuner. In addition, the tuner output audio might damage the microphone.

An *audio mixer* eliminates all of the problems involved with the connection of multiple devices to a single channel. First, it isolates the inputs from each other, so there is no impedance mismatch or competition among the sources. Second, the gain (amplification factor) at each input can be varied independently. This allows adjustment of amplitudes so the signals blend in the desired ratio.

FILTERS AND EQUALIZERS

Audio filters and *equalizers* allow for the adjustment of loudness of audio signals at various frequencies. They are used in recording studios and by serious hi-fi stereo enthusiasts, and are discussed in Chapter 11.

PROBLEM 7-4

In the tone control arrangement shown by Fig. 7-3B, what would happen if the capacitor in the *RC* circuit labeled Y were to short out?

SOLUTION 7-4

This would be the equivalent of setting potentiometer Y at its minimum resistance all the time, so the bass audio level control would stop working.

Amplifier Considerations

In hi-fi, an amplifier delivers medium or high audio power to a set of speakers. There is at least one input, but more often there are three or more: one for a CD player, another for a tuner, and still others for auxiliary devices such as a tape player, turntable, or computer. Input requirements are a few milliwatts; the output can range up to hundreds of watts.

LINEARITY

Linearity is the extent to which the output waveform of an amplifier is a faithful reproduction of the input waveform. In hi-fi equipment, all amplifiers must be as *linear* as possible. Linearity in an amplifier can be thought of as the equivalent of optical quality in a microscope or telescope.

If you connect a *dual-trace oscilloscope* (one that lets you observe two waveforms at the same time) to the input and output terminals of a hi-fi audio amplifier with good linearity, the output waveform is a vertically magnified duplicate of the input waveform. When the input signal is applied to the horizontal scope input and the output signal is applied to the vertical scope input, the display is a straight line. In an amplifier with poor linearity, the instantaneous output-versus-input function is not a straight line, indicating that the output waveform is not a faithful reproduction of the input. In this situation, *distortion* occurs. Sometimes listeners cannot notice distortion, even if it shows up clearly on an oscilloscope. But even if it is not directly perceived, it can have a psychological effect.

Hi-fi amplifiers are designed to work with input signals up to a certain peak (maximum instantaneous) amplitude. If the peak input exceeds this level, the amplifier becomes nonlinear, and distortion is inevitable. In a hi-fi system equipped with *volume-unit meters* (also called VU meters) or *distortion meters*, excessive input causes the needles to "kick up" into the red range of the scales during audio peaks. This condition should be avoided.

DYNAMIC RANGE

Dynamic range is a prime consideration in hi-fi recording and reproduction. As the dynamic range increases, the sound quality improves for music or programming having a wide range of volume levels. Dynamic range is the difference between the strongest and the weakest output audio that a system can produce without objectionable distortion taking place. It is usually specified in decibels (dB) based on sound power.

At low volume levels, the limiting factor in dynamic range is the *background noise* in the system. In an analog system, most of this noise comes from the audio amplification stages. In tape recording, there is also some *tape hiss*. A scheme called *Dolby* (a trademark of Dolby Laboratories) is used in professional recording studios, and also in high-end consumer tape equipment, to minimize this hiss. Digital recording systems produce less internal noise than analog systems. For this reason, digital systems have superior dynamic range.

At high volume levels, the power-handling capability of an audio amplifier limits the dynamic range. If all other factors are equal, a 100-W audio system can be expected to have greater dynamic range than a 50-W system. The speaker size is also important. As speakers get physically larger, their ability to handle high power improves, resulting in increased dynamic range. This is why serious audio enthusiasts sometimes purchase sound systems with amplifiers and speakers that seem unnecessarily large. An "overengineered" system can sometimes be a symptom of megalomania. But there is some common sense in it too, because it ensures that occasional extreme audio peaks can be reproduced without distortion.

Mobile and Portable Systems

Hi-fi systems designed for use in cars and trucks, or for private listening with headphones, operate at low DC voltages. Audio power levels are much lower than in home hi-fi systems. Most such systems operate using batteries rather than a power supply connected to the electric utility mains. However, all of these systems have two stereo channels, just as does the largest home hi-fi system.

MOBILE SYSTEMS

A *mobile hi-fi system*, designed for a car or truck, typically has four speakers. The left and right channels each supply a pair of speakers. The left stereo channel drives the left front and left rear speakers; the right stereo channel drives the right front and right rear speakers. The balance control adjusts the ratio of sound volume between the left and right channels for both the front and rear speaker sets. Another control adjusts the ratio of sound volume between the front and rear sets.

A mobile hi-fi system has an AM/FM receiver and a CD player. Some older vehicles have systems with cassette tape players. Newer, high-end cars and trucks have satellite radio receivers as well as conventional AM/FM receivers and CD players. One note of caution: CDs and tape cassettes are heat-sensitive, so they should not be stored in a car or truck that will be left out in the sun on a warm day.

PORTABLE SYSTEMS

A *portable hi-fi system* operates from dry cells or rechargeable cells. The most well-known type is the so-called *headphone radio*. There are dozens of different designs available. Some include only an FM radio; some have AM/FM reception capability. Some have a small box with a cord that runs to the headset; others are entirely contained within the headset. There are portable CD players and even portable satellite radio receivers. The sound quality from these systems can be excellent, although some are mediocre, especially those that lack tone controls.

Another form of portable hi-fi set, sometimes called a *boom box*, can produce several watts of audio output, and delivers the sound to a pair of speakers built into the box. A typical boom box is about 20 cm (8 in) high by 40 cm (16 in) wide by 15 cm (6 in) deep. It includes an AM/FM radio and a CD player. The system gets its name from the loud bass acoustic energy peaks its speakers can deliver. The level of this bass audio can be astonishing, considering the fact that the speakers are comparatively small!

PROBLEM 7-5

I bought a tiny FM portable radio to wear while walking around the track at the health club. The thing is so small that I can hardly tell it's there when I wear it with the armband provided. However, when I use the headset from my home hi-fi system with it, this little radio sounds as good as the FM tuner in my expensive home audio system, although I can't get the volume up as loud. How is this possible? Even the bass sounds great.

SOLUTION 7-5

Electronic circuits can amplify signals at all frequencies "from DC to light," and with modern IC technology, the signal quality does not depend on the physical size of the circuit or on the frequency. Bulky electronics do not guarantee good sound, and microminiature electronics do not preclude it. In your case, it's the high-quality headset you're using with the little radio that makes the difference. Try plugging in a cheap headset (like the one that might have come with the little portable radio) into your home audio system! The sound quality in that case will be just as bad with your expensive amplifier as it is with the little radio, or with any other radio.

Quiz

Refer to the text in this chapter if necessary. A good score is 8 correct. Answers are in the back of the book.

1. A midrange musical note
   (a) has a frequency exactly in the middle of the audible range.
   (b) has volume neither too loud nor too soft.
   (c) has a frequency higher than bass notes but lower than treble notes.
   (d) None of the above

2. In the acoustical design of a room intended for a home audio system,
   (a) the use of small speakers can minimize distortion.
   (b) reflection of sound waves from walls should be minimized.
   (c) the walls should all intersect at perfect 90° angles.
   (d) wooden furniture, without upholstery, should be used.

3. A change of +10 dB in an audio signal represents
   (a) a doubling of acoustic power.
   (b) an increase in acoustic power by a factor of 3.
   (c) an increase in acoustic power by a factor of 10.
   (d) an increase in frequency by a factor of 10.

4. A sound wave that propagates at 335 m/s in dry air at sea level has a frequency of
   (a) 33.5 Hz.
   (b) 335 Hz.
   (c) 3.35 kHz.
   (d) Hold it! A sound wave in dry air propagates at 335 m/s no matter what its frequency. So the frequency could be anywhere in the audible range.

5. The relative phase of two acoustic waves from the same source at the same time, one wave direct and one wave reflected from a wall, can affect
   (a) the locations of antinodes and nodes.
   (b) the perceived frequency.
   (c) the locations of antinodes and nodes, and the perceived frequency.
   (d) neither the locations of the antinodes and nodes, nor the perceived frequency.

6. In an acoustic sine wave,
   (a) the frequency and phase are identical.
   (b) the sound power is inversely proportional to the frequency.
   (c) the sound power is directly proportional to the frequency.
   (d) all of the sound power is concentrated at a single frequency.

7. Which of the following frequencies cannot be received by an AM/FM tuner?
   (a) 830 kHz.
   (b) 95.7 kHz.
   (c) 90.1 MHz.
   (d) 107.3 MHz.

8. Which of the following statements about woofers is true?
   (a) They are especially useful for reproducing the sounds of barking dogs.
   (b) They are designed to handle short, intense bursts of sound.
   (c) They should not be used with tone controls.
   (d) They are specifically designed to reproduce low-frequency sounds.

9. A rack-mounted hi-fi system
   (a) can save floor space.
   (b) is more susceptible to distortion than a compact system.
   (c) is cheaper than a compact system.
   (d) is designed especially for use with headsets.

10. An audio mixer
    (a) cannot match impedances among interconnected components.
    (b) cannot, by itself, increase the audio output of an amplifier.
    (c) can eliminate distortion, no matter what else is wrong with a system.
    (d) allows a microphone to be used as a speaker.

Speakers, Headsets, and Microphones

The fidelity of sound reproduced by an audio system depends on the quality of the speakers or headset. The integrity of recorded or live audio is a function of the quality of the microphones. No sound system can be better than its *transducers*—the devices that convert electrical signals to acoustic waves, or vice-versa. In this chapter, you will learn how speakers, headsets, and microphones work.

Common Speaker Types

All speakers are *electro-acoustic transducers*. They convert electrical energy in the form of AF signals to acoustical energy in the form of moving air molecules (sound waves).

DYNAMIC SPEAKER

Figure 8-1 is a simplified functional diagram of a *dynamic speaker*. The transducer works on the basis of interaction between the constant magnetic field from a permanent magnet and the varying magnetic field produced by a coil that is supplied with the AF current output from an amplifier.

In the basic dynamic transducer design, a *diaphragm* is attached to a coil that is mounted so it can move freely and rapidly, but with a small displacement, back and forth along its axis. The coil surrounds a permanent magnet. When an AF current is applied to the coil, it generates an alternating magnetic field around the coil. This causes the coil to move because of varying magnetic forces between the coil and the magnet. (Some dynamic speakers use DC electromagnets, rather than permanent magnets, to produce stationary magnetic fields.) The movable coil is rigidly attached to the large diaphragm, so the coil motion causes an acoustic disturbance as the diaphragm exerts alternating in-and-out pressure on the surrounding air.

In a conventional speaker, the coil is called the *voice coil*, the diaphragm is called the *cone*, and the cone is attached at its periphery to a fixed *suspension*, also known as a *spider* because of its shape. The whole speaker is usually contained in a *cabinet* with an open-air front. A *grille* can be placed in front of the speaker to protect the speaker cone from the external environment. Internal *baffles* can be used to tailor the frequency response of the speaker and its enclosure.

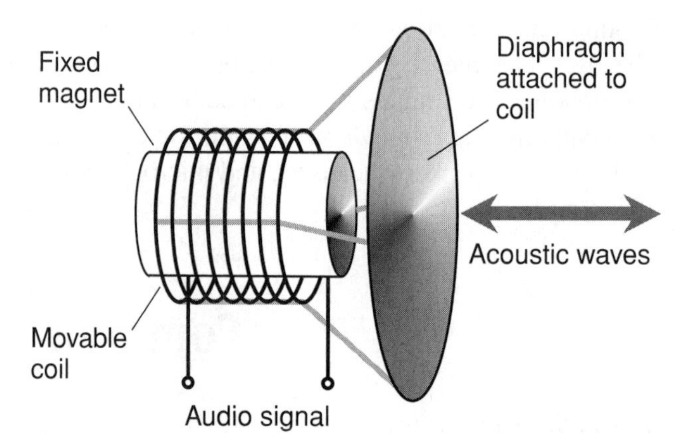

Figure 8-1 Functional diagram of a dynamic speaker. Most speakers operate according to this principle, although geometries vary greatly.

ELECTROSTATIC SPEAKER

An *electrostatic speaker* employs electric fields rather than magnetic fields to produce the forces that vibrate its moving element. A typical electrostatic speaker has a fixed, rigid, flat metal sheet and a flexible, flat metal sheet (Fig. 8-2).

The AF signal from an amplifier is applied to the terminals of the transducer, first passing through a step-up transformer, then through a *blocking capacitor* that prevents the DC source from being shorted out through the transformer secondary, and finally onto the metal plates. The AF voltage, combined with the DC source voltage, creates a fluctuating DC voltage between the plates. The result is an *electrostatic field* of constant polarity but varying intensity. This exerts a fluctuating force between the plates, causing the flexible plate to vibrate and generate acoustic waves in the air.

An alternative electrostatic speaker design has a flexible metal sheet placed in between two rigid sheets. The rigid sheets are perforated to allow sound waves, produced by the flexible sheet as it moves, to escape from the speaker into the air.

Electrostatic speakers are comparatively expensive. They can be recognized by their flat, thin structure, compared with the "boxy" shape of most dynamic speaker cabinets. Electrostatic speakers can respond well to a wide range of frequencies, so that a single unit can generally handle the bass, midrange, and treble sounds. Electrostatic speakers have limited ability to handle high audio power, and are easily damaged if overdriven.

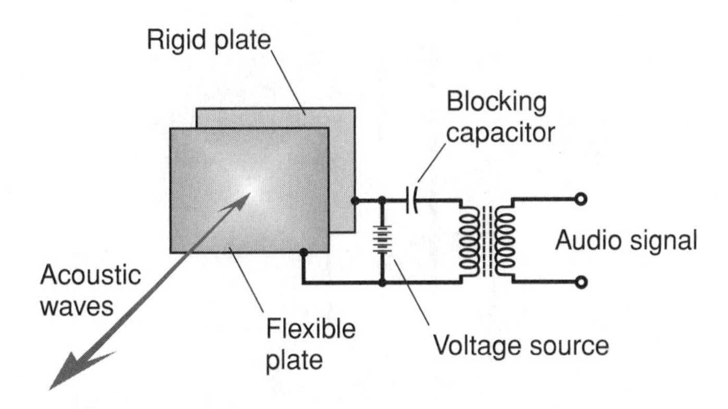

Figure 8-2 Functional diagram of an electrostatic speaker.

PIEZOELECTRIC TRANSDUCER

Figure 8-3 is a cross-sectional view of a *piezoelectric transducer*. This device consists of a *crystal* of quartz or ceramic material sandwiched between two metal plates. If an AF voltage is applied to the plates, it causes the crystal to vibrate in sync with the fluctuating voltage between the plates. The metal plates vibrate because they are in direct contact with the crystal. This causes an acoustic disturbance in the surrounding air.

Piezoelectric transducers can function at higher frequencies than can dynamic or electrostatic devices. For this reason, they are favored in applications involving *ultrasound*, such as intrusion detectors, motion detectors, and robotic navigation systems. However, piezoelectric transducers do not work well at the lower midrange and bass frequencies. They are physically fragile. If a piezoelectric transducer is accidentally struck or dropped, the crystal may fracture, permanently destroying the device.

WOOFERS, MIDRANGE SPEAKERS, AND TWEETERS

In general, as the minimum frequency of the AF signal to be converted into sound goes down, the optimum diameter of a speaker cone increases. Conversely, as the maximum frequency rises, the optimum cone diameter decreases. This is because

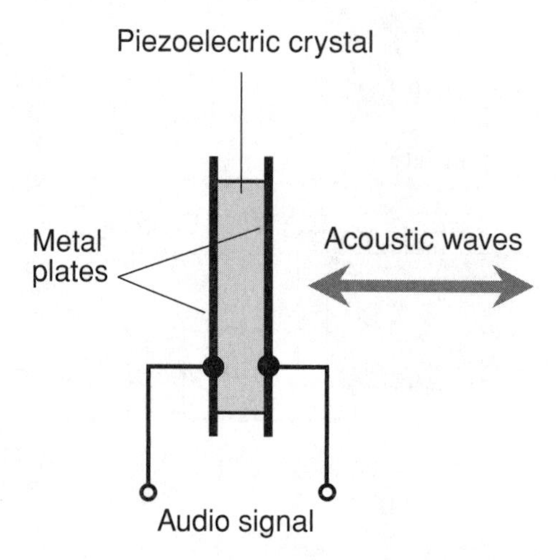

Figure 8-3 Functional diagram of a piezoelectric transducer.

low audio frequencies have long wavelengths, requiring large cones that can displace large quantities of air, while high frequencies require small cones that can vibrate rapidly.

High-end dynamic speaker systems in high-fidelity (hi-fi) equipment usually consist of three separate transducers for optimum coverage of the AF spectrum. A *woofer* is a large speaker designed for the bass frequencies below about 200 Hz. A *midrange speaker* is smaller and handles AF between about 200 Hz and 3 kHz. A *tweeter* handles the treble frequencies above 3 kHz. In some systems, there are only two speakers—one for low frequencies and the other for midrange and high frequencies—with some overlap in their response range.

In voice communications, speakers are almost always of the dynamic type, and are in effect miniature midrange speakers. The power requirements are modest. High fidelity is not as important as *intelligibility*, the ease with which the voice output is understood. A typical communications speaker measures 5 to 10 cm (2 to 4 in) across. Speakers intended for computer use are similar to communications speakers. They are sometimes built into the computer's main unit or external display. Such speakers usually contain small internal audio amplifiers that require dry cells to function. Some people connect the audio outputs of their computers to hi-fi amplifier/speaker systems to obtain enhanced sound.

CROSSOVER NETWORKS

A *crossover network* is needed for proper operation of a dynamic speaker combination. The woofer should receive mostly bass audio; the midrange unit should get primarily midrange audio; the tweeter should receive mainly treble audio. It is especially important that the midrange and tweeter units not be subjected to strong bass signals, because this can result in permanent physical damage.

The most basic form of crossover network consists of a circuit called a *lowpass filter* between the amplifier output and the woofer, and another circuit called a *highpass filter* between the amplifier output and the tweeter and midrange speaker. Figure 8-4 shows schematic diagrams of a simple lowpass filter (at A) and a simple highpass filter (at B). These are inductance-capacitance (*LC*) circuits. Because of the way they look in schematic diagrams, the arrangement at A is called a *pi network* and the arrangement at B is called a *T network*. The optimum component values must be found by experimentation.

Figure 8-5 illustrates a generic lowpass frequency response curve (at A) and a generic highpass frequency response curve (at B). In some systems, two separate highpass filters are used, one for the tweeter and another for the midrange speaker.

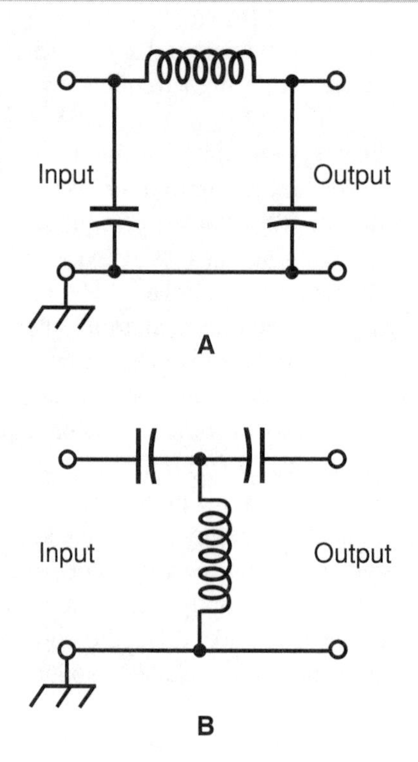

Figure 8-4 At A, a simple lowpass filter circuit. At B, a simple highpass filter circuit.

PROBLEM 8-1

Is there a simpler way to make a crossover network than the use of the lowpass and highpass filters shown in Fig. 8-4?

SOLUTION 8-1

Yes, although it represents a compromise. The simplest lowpass filter is a capacitor in parallel with the leads to a woofer. The simplest highpass filter is a capacitor in series with the live (signal-carrying) lead to a tweeter or midrange speaker. The response curves with single capacitors are not as well-defined as the curves shown in Fig. 8-5. There is some interaction between the capacitors, and their values must be chosen by experimentation. Figure 8-6 is a simple diagram of how two capacitors can be connected to make a *first-order crossover circuit* for a speaker system consisting of a woofer and a tweeter only. This scheme can sometimes prove satisfactory in modest audio systems.

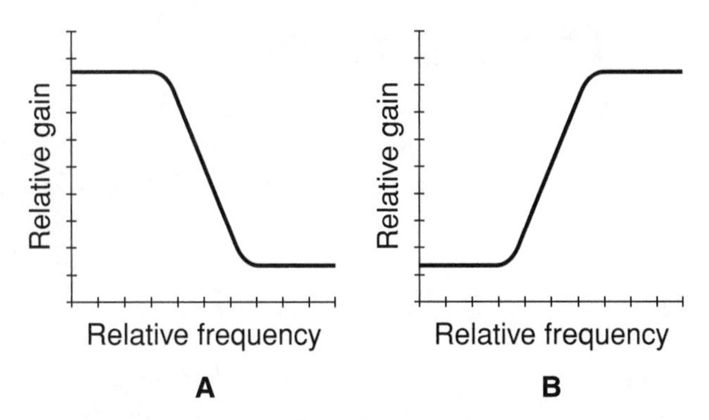

Figure 8-5 At A, a lowpass frequency response curve. At B, a highpass frequency response curve.

PROBLEM 8-2

What is the function of the *impedance-matching transformer* in Fig. 8-6?

SOLUTION 8-2

This component optimizes the transfer of power from the amplifier output to the speakers. It is a transformer with a powdered-iron core, designed especially to work at AF. *Impedance* is a quantitative expression of the extent to which a component opposes any form of AC, including utility power, AF current, and RF current. Speaker impedance is defined in the next section. Impedance is explained in more detail in Chapter 9.

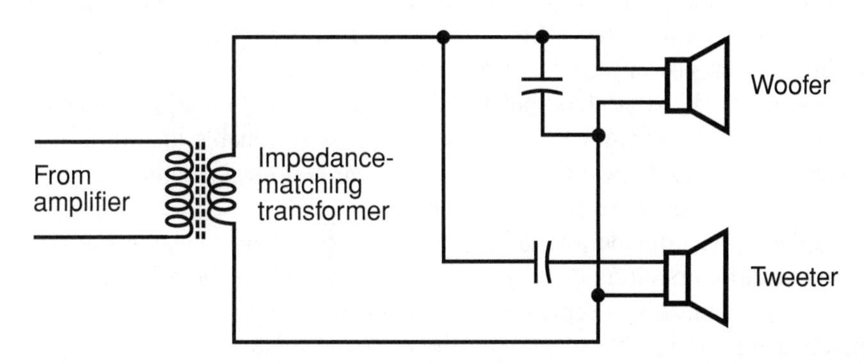

Figure 8-6 Illustration for Problems and Solutions 8-1 and 8-2.

General Speaker Considerations

When choosing a set of speakers for an audio system, several factors must be considered. Here are some of the specifications most often mentioned.

PHYSICAL SIZE AND APPEARANCE

The physical appearance of a set of hi-fi speakers is largely a matter of user preference, but there are also some practical concerns. Huge speakers and powerful audio systems will not sound very good in small rooms. Even the slightest hum or extraneous noise might be audible. At high volume levels, the room will be "acoustically overwhelmed." Conversely, tiny speakers are inadequate for use in large rooms, at parties, in classrooms, or in auditoriums for obvious reasons.

Some of the most serious audiophiles regard speakers as a personality statement. For example, the absence of a speaker grille (covering) exposes the cones and/or horns, and can convey a sense of raw power. It also increases the vulnerability of speakers to physical damage from spilled liquids, inquisitive toddlers, and free-running pets.

POWER-HANDLING CAPACITY

Speakers must handle the audio peaks at all frequencies the amplifier delivers. Some amplifiers produce high-power peaks at frequencies below the human hearing range; speakers must deal with such signals effectively. If you have an amplifier rated at 50 W peak output, for example, then the speakers must be able to withstand at least 50 W peak throughout the frequency range of the amplifier.

It's a good idea to allow a liberal margin of safety when choosing speakers according to their power rating. For example, it is wise to purchase 100-W speakers for use with a 50-W peak output amplifier (a 3-dB safety margin). Intense low-frequency or high-frequency sound bursts cause objectionable distortion in small speakers, and can ruin speakers that are not large and massive enough to handle the physical vibration these peaks can cause.

With the above considerations in mind, it doesn't make sense to use gigantic speakers with tuners alone or with tiny amplifiers. While there is nothing technically wrong with this, it represents overkill and is a waste of money. But such extravagance is preferable to making the opposite mistake, and using speakers that are not large enough to handle the audio output power from the amplifier system.

LOW POWER AND TWEETERS

There's an exception to the above rule when low-power amplifiers are used with tweeters. If a low-power amplifier is run at a volume setting at or near its maximum, the amplifier can produce high levels of audio energy at harmonic frequencies. Some of this energy is beyond the range of human hearing, but its peaks can attain levels sufficient to damage small tweeters, even if they are rated at a power level greater than that of the amplifier. Low-power amplifiers should not be operated near or at their highest volume settings for this reason. If you want loud sound, get a big amplifier and big speakers!

PEAK VERSUS RMS POWER

Another fact worth noting is that speakers are often rated for the *rms power* they can handle, and not the *peak power*. As you learned in Chapter 2, rms values are generally less than peak values. It is sometimes possible to use a speaker whose rms power rating is the same as, or even a little less than, the peak power rating of the amplifier, as long as the amplifier is not operated at a volume level high enough to cause distortion. The best way to prevent confusion in this respect is to be sure that the rms power rating for one component is not compared with the peak power rating for another component.

When buying an audio system, it is a good idea to discuss both the amplifier and speaker specifications with a knowledgeable technician or engineer. Then you can be sure the speakers are well matched to the amplifier system, electronically as well as esthetically.

SPEAKER IMPEDANCE

Impedance is a measure of the extent to which an electronic component, device, or system opposes the flow of AC. It is the AC equivalent of DC resistance. Like DC resistance, impedance is expressed in ohms (Ω). When it appears as a variable in an equation, impedance is represented by the uppercase, italic letter Z.

Impedance is more complex than DC resistance, because it involves inductive and capacitive effects as well as ordinary resistance. The impedance of any device that contains inductance or capacitance, or both, varies with frequency. Speakers are no exception, although engineers strive to build speakers that exhibit constant impedance over the entire AF range. Dynamic speakers nearly always have an impedance peak somewhere in the bass range, and a generally rising impedance as the frequency increases through midrange, treble, and ultrasonic frequencies. Most speakers are designed to have nominal impedances of 8 Ω, but some are rated at 4 Ω.

If you measure the DC resistance of a dynamic speaker coil with an ohmmeter, you should expect to get an ohmic value less than the rated impedance of the speaker. This is because the inductance of the coil offers no resistance to DC; the only contributor to pure resistance is *ohmic loss* in the coil wire. However, when AF current passes through the coil, the coil alternately stores and releases energy in the form of a magnetic field. This effect is manifested as an effective increase in the AC resistance.

Electrostatic speakers have inherently high impedance, because their plates are not directly connected. However, this high impedance can be matched to the output of an amplifier by means of a transformer (such as is shown in Fig. 8-2), so the amplifier "sees" an impedance similar to that provided by a dynamic speaker.

As a fundamental rule, the impedance that a speaker (or set of speakers) presents to the output of an audio amplifier should never be less than 4 Ω. That means you can connect two 8-Ω speakers in parallel, but you should not connect a 4-Ω speaker in parallel with any other transducer.

FREQUENCY RESPONSE

A theoretically perfect speaker exhibits a *flat frequency response* over the entire range of audio frequencies, and even somewhat above and below that range, with pure sine-wave signal input. That means the actual output sound power is in the same proportion to the input signal power at all frequencies. This level of perfection is never achieved in practice, although, with carefully designed combinations of woofers, midrange speakers, and tweeters, it can be approached. A good dynamic hi-fi speaker cabinet is free of sharp *acoustic resonances* that can cause hollow or tunnel-like sound output. However, broad resonances at low frequencies can enhance the bass performance of small speakers.

The geometry of the speaker cabinet, and the material from which it is made, affects the way dynamic speakers perform in the real world. Two identical speakers in different cabinets will produce dramatically different sounds. There are numerous ways in which speaker cabinets can be designed for optimum performance. Unfortunately, there are even more ways in which speakers cabinets can be poorly designed, resulting in marginal speaker performance. When buying speakers, it is a good idea to listen to several models and compare their performance in an environment similar to that where they will be used.

Amplifier tone controls and graphic equalizers can compensate for poor speakers to some extent, but they should never be relied upon for that purpose.

DISPERSION

In audio practice, *dispersion* is an expression of the extent to which a speaker emits acoustic waves in all directions. The best possible dispersion—a theoretical ideal—is an *omnidirectional sound radiation pattern.* In practical scenarios, a speaker can be considered to have good dispersion if it radiates sound over an angle wide enough so that all listeners in the environment hear the sound at the same volume when they are equidistant from the speaker. This is based on the assumption that there are no major flaws or anomalies in the acoustic characteristics of the room where the speaker is located.

Speakers become more directional—that is, the dispersion decreases—as the frequency increases. The effect is gradual up to a certain frequency, but above that frequency it becomes dramatic. As a general rule, a speaker can be expected to have good dispersion if its cone diameter is less than one wavelength (1 λ). Above the frequency at which the cone diameter is 1 λ, the dispersion deteriorates.

The highest frequency at which a speaker cone of a given diameter can be expected to exhibit good dispersion can be calculated. Suppose that d_{cm} is the cone diameter, measured in centimeters. Then the highest frequency f_{Hz}, in hertz (Hz), for which good dispersion can be expected is given by:

$$f_{Hz} = 33{,}500/d_{cm}$$

If d_{in} is the cone diameter in inches, then:

$$f_{Hz} = 13{,}200/d_{in}$$

Sometimes the audio frequency is expressed in kilohertz (kHz) rather than in hertz. In that case, if we let the frequency be symbolized f_{kHz}, the above formulas become:

$$f_{kHz} = 33.5/d_{cm}$$

and

$$f_{kHz} = 13.2/d_{in}$$

From this, it is evident that if good dispersion is desired at high frequencies, small speakers are required. But small speakers don't reproduce bass very well. That presents a conundrum! A common solution is the use of a woofer for the bass, typically having a diameter of 15 cm (6 in) or more, and a tweeter for the treble, having a diameter of around 2.5 cm (1 in) or less. A midrange speaker can be included to ensure that there is no performance gap where either the dispersion or the sound fidelity are compromised. If single-transducer speakers (as opposed to woofer/tweeter or woofer/midrange/tweeter composites) are desired for economic reasons, moderately large units are available with tweeter-like secondary transducers at the centers of their cones. Such a transducer is called a *whizzer*. Like a tweeter, the whizzer has a diameter of about 2.5 cm (1 in) or less. It can be recognized by its dome-like shape.

PROBLEM 8-3

Suppose the diameter of a woofer cone is one foot (1 ft). Also suppose this woofer has no whizzer for reproduction of treble. What is the highest frequency at which this woofer can be expected to exhibit good dispersion characteristics? Express the answer in hertz, and also in kilohertz.

SOLUTION 8-3

Use the second formula above, noting that 1 ft = 12 in. Calculate as follows:

$$\begin{aligned} f &= 13{,}200/d_{\text{in}} \\ &= 13{,}200/12 \\ &= 1100 \text{ Hz} \\ &= 1.1 \text{ kHz} \end{aligned}$$

TRANSIENT RESPONSE

The *transient response* of a speaker is a qualitative measure of its ability to deal with brief, high-intensity audio signal bursts or "peaks." An ideal speaker responds immediately to such bursts, and in proportion to their intensity. Also, a theoretically perfect speaker cone stops moving (or resumes its normal motions in proportion to the audio signals the speaker receives) immediately after the transient ends.

In the real world, speakers are a little slow to respond to sudden bursts of audio. The cone has a certain amount of *inertia* because of its physical mass. Also because of inertia, a real-world speaker cone takes a small amount of time to come to rest after an audio peak has passed, a phenomenon called *hangover*. The response and hangover times should be as short as possible. The longer they are, the "mushier" the reproduced sound will be, especially for music that contains intense and dramatic sound peaks.

In general, the best speaker in regards to transient response is one with a nearly flat frequency response, and in particular, one that is devoid of significant acoustic resonances, with the exception of the broad resonance in the bass range that is a normal characteristic of dynamic speakers. Resonances tend to worsen the transient response of any speaker.

EFFICIENCY

Speaker efficiency is the ratio of the actual sound power output to the AF power applied to a speaker. If P_{in} is the AF input power to a speaker (in watts) and P_{out} is the sound power it produces (also in watts) as a result of the input signal, then the efficiency, *Eff*, is:

$$Eff = P_{out}/P_{in}$$

As a percentage, the efficiency $Eff_{\%}$ is:

$$Eff_{\%} = 100\ P_{out}/P_{in}$$

The efficiency of a speaker depends on the waveform applied to it, the frequency (or component frequencies) of the wave, and the actual power input. Most real-world speakers have a single frequency at which their efficiency is maximum. A theoretically perfect speaker would have the same efficiency at all frequencies in the AF spectrum.

Speakers are not efficient transducers. This is especially true of dynamic speakers in enclosed cabinets, which sometimes have efficiency figures of less than 1 percent. This means, for example, that a 10-W audio signal applied to such a speaker will produce less than 100 mW of actual sound power in the surrounding air! Open-air speakers, especially larger ones incorporating woofers, midrange transducers, and tweeters, can have efficiency ratings as high as about 10 percent.

Efficiency is not especially important to most audio enthusiasts. Even a speaker that is 0.5 percent efficient can, with a reasonable amplifier, produce more than enough sound to satisfy the demands of ordinary hi-fi home users. The fidelity of the sound is of much greater concern. This is a function of the frequency response, the dispersion, and the transient response.

Headsets

A *headset* consists of a flexible, elastic *headband* and two small transducers called *earphones*. The earphones in most headsets are dynamic transducers, and resemble pairs of miniature loudspeakers. One earphone reproduces left-channel sound, and the other earphone reproduces right-channel sound. Headsets can render high-fidelity stereo sound with vivid channel separation. They can also provide excellent intelligibility in communications.

Headsets are available with impedance ratings from a few ohms to more than 2,000 Ω. Some headsets have nearly flat frequency responses. These are best for high-fidelity applications. Other types have peaked responses. These are intended for communications use, particularly radio communications in which noise and interference can be problematic. The proper impedance and frequency response should be chosen for the intended application. If you are not sure about the optimum impedance for a headset, check the instruction manual of the device or system with which you intend to use it.

Many types of headsets are available, ranging in price from a few dollars to several thousand dollars. Headsets of reasonable quality can be bought in any electronics store where high-fidelity and home-entertainment equipment is sold. Some headsets have miniature radio receivers built-in. When purchasing a headset for stereo high-fidelity use, it's important to listen to several different models while in the store. It is amazing how much difference there is in the way various models perform. To some extent, these differences can be compensated for by means of tone controls and/or a graphic equalizer, but those devices cannot make a bad headset sound like a good one. Your headset should be comfortable, and should not put excessive pressure on your ears or on the skin around your ears. Otherwise, you'll experience "headset fatigue" after long periods of listening.

A headset with both earphones wired directly together in parallel or in series is called a *monaural headset*. These headsets cannot reproduce stereo sound. They are used almost exclusively in communications. Some monaural headsets have a *boom* attached with a small microphone at the end. These can be used for hands-free communicating by radio or by telephone.

Headsets can impair the listener's ability to hear external sounds. This is true even of single-earphone designs, but it applies especially to two-earphone types. In some states and municipalities, it is illegal to drive a motor vehicle while wearing a headset.

PROBLEM 8-4

Can a headset be directly connected to the speaker output terminals of an amplifier, in parallel with the speakers themselves? When I plug the headset into the jack provided on the panel of my amplifier, the speakers go silent. I want sound to come from the speakers and the headset at the same time.

SOLUTION 8-4

This can be done, but resistors should be installed in series with each earphone to ensure that the headset does not receive an excessively strong signal. These resistors (the values of which should be determined by experimentation) will also prevent the load impedance from becoming too low. Figure 8-7 shows the basic scheme. The resistors should be in the live earphone leads, and run to the red speaker terminals or to the center conductor receptacles of the speaker jacks. The normally grounded lead(s) from the headset should be connected directly to the

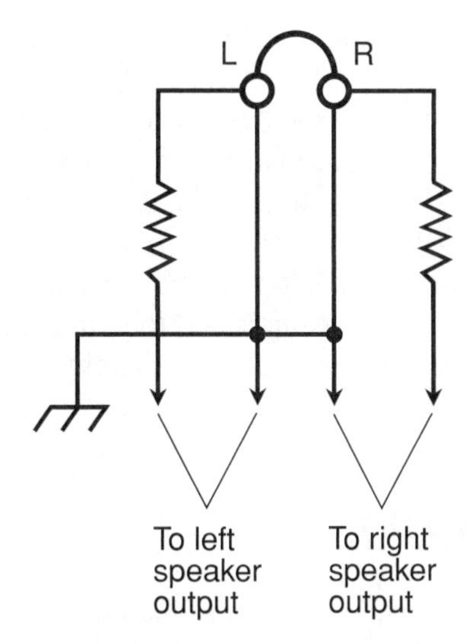

Figure 8-7 Illustration for Problem and Solution 8-4.

black speaker terminals, or to the grounded outer shells of the speaker jacks. (These polarity considerations ensure that neither speaker output will be short-circuited.) Start with resistors of about 100 times the headphone impedance (800-Ω resistors for an 8-Ω headset, for example), and decrease the resistances until the desired volume characteristics are achieved. Potentiometers, in series with fixed resistors of the same ohmic value as the headset impedance, can be used to provide continuously adjustable gain for each channel, but care must be taken to be sure the potentiometer resistances are not set too low.

Microphones

A *microphone* is an electro-acoustic transducer specifically designed to convert sound waves into AF electrical signals. The most common microphones are, in effect, miniature speakers that work in reverse.

DYNAMIC MICROPHONE

A *dynamic microphone* takes advantage of magnetic fields and mechanical movements to generate AF current from acoustic waves. The principle of operation is exactly the same as that of the dynamic speaker, except that it operates in the opposite sense.

For reference, see the functional diagram of Fig. 8-1. Vibrating air molecules cause mechanical movement of the diaphragm. The coil, attached rigidly to the diaphragm, moves along with it in the field of the permanent magnet. As a result of this movement, currents are induced in the coil. These currents are in exact synchronization with the sound waves striking the diaphragm.

The dynamic microphone is the most common type used today. All dynamic microphones are physically and electrically rugged, and last a long time. However, the magnet should not be subjected to extreme heat or repeated physical shock, because these abuses can weaken it, thereby reducing the sensitivity of the microphone.

ELECTROSTATIC MICROPHONE

An *electrostatic microphone* is, in effect, a miniature electrostatic speaker "working backwards." For reference, see the functional diagram of Fig. 8-2. When a sound wave strikes the flexible plate, the plate vibrates. This causes rapid changes in the spacing between the flexible plate and the rigid plate, producing fluctuation in the capacitance between the plates. A DC voltage source is connected to the plates. As the capacitance increases and decreases, the plates alternately charge and discharge. This causes a weak AC signal to be produced, whose waveform is similar to that of the acoustic disturbance. The blocking capacitor allows this AC signal to pass to the transducer output, while keeping the DC confined to the plates. The transformer matches the output impedance of the microphone to the input impedance of an audio amplifier.

CRYSTAL AND CERAMIC MICROPHONES

A *crystal microphone* takes advantage of the *piezoelectric effect* to convert sound waves into electrical impulses. For reference, see the functional diagram of Fig. 8-3. Vibrating air molecules set the metal plate or plates in motion. The plates, connected physically to the crystal, put mechanical stress on the piezoelectric substance. This results in small electrical currents that vary at the same frequency or frequencies as the sound.

Ceramic is a manufactured compound consisting of aluminum oxide, magnesium oxide, or other similar materials. It is a white, fairly lightweight solid. When subjected to the stresses of mechanical vibration, certain ceramic materials generate electrical impulses. A *ceramic microphone* uses a ceramic cartridge to transform sound energy into an AF signal. Its construction is similar to that of a crystal microphone.

Crystal and ceramic microphones have high output impedance, and respond well to high-frequency sound and ultrasound.

DIRECTIONAL MICROPHONE

A *directional microphone* is designed to be more sensitive in some directions than in others. Usually, a directional microphone is *unidirectional* — that is, its maximum sensitivity occurs in only one direction.

Directional microphones are used in all types of audio systems to reduce the level of background noise that is picked up. This is important for intelligibility, especially in environments where the ambient noise level is high. Directional microphones are also useful in *public-address* (PA) *systems* because they minimize the amount of feedback from the speakers. Directional microphones can usually be recognized by their physical asymmetry.

Most directional microphones have a *cardioid directional response*. The zone of maximum sensitivity of a *cardioid microphone* exists in a very broad *lobe*. The direction of minimum sensitivity is sharply defined, and occurs directly opposite the direction of greatest sensitivity.

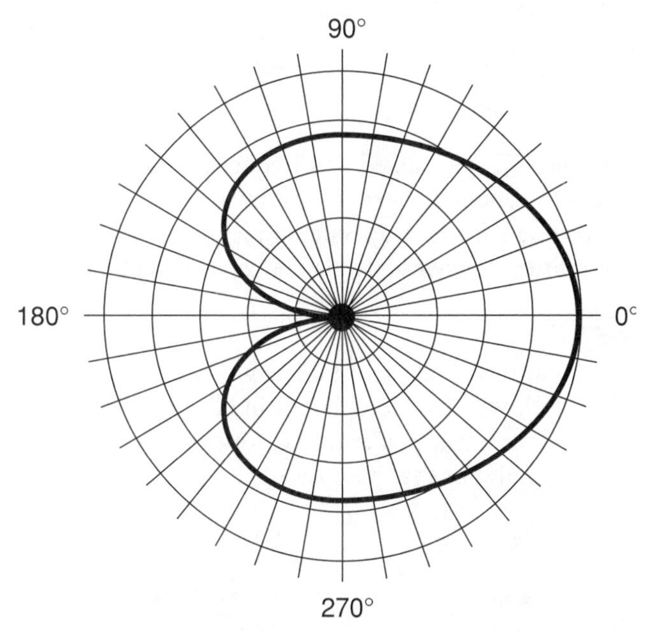

Figure 8-8 Directional plot for a cardioid microphone. Maximum sensitivity is from 0° and minimum sensitivity is from 180°. The microphone is at the center of the coordinate system.

Figure 8-8 is a *directional plot* of a cardioid response. The response exhibits *axial symmetry*, meaning that the curve looks the same (resembling Fig. 8-8) in all planes passing through the head of the microphone and containing the line that defines the direction of maximum sensitivity. Some directional microphones have sharper responses than that shown in the figure, so they are less responsive to sounds coming from the sides. Some have a deeper null than others in the direction of minimum sensitivity.

OMNIDIRECTIONAL MICROPHONE

An *omnidirectional microphone* is equally sensitive to sound from all directions. Microphones of this type work well when the ambient noise level is low, or when it is necessary to pick up sound from widely separated sources. An application where omnidirectional microphones are ideal is the recording of a symphony orchestra as heard from the conductor's position.

In general, if sound comes from a number of different directions, or if the source is moving with respect to a listener, omnidirectional microphones are preferable for recording. Omnidirectional microphones typically have spherical heads that may be covered with foam padding to reduce unwanted pickup of sounds caused by a person's breath in voice recording or by a light wind in outdoor recording.

PROBLEM 8-5

Can a woofer operate as a specialized low-frequency microphone? Can a tweeter be used as a specialized high-frequency or ultrasonic microphone? Can a directional midrange horn be used as a directional pickup — say, in conjunction with sensitive audio amplifiers to eavesdrop on conversations at a distance?

SOLUTION 8-5

Theoretically, the answer to all these questions is "Yes." However, you should not expect speakers to work well as microphones in hi-fi applications. If you want to make a recording of a live music performance, for example, you should use microphones designed especially for that purpose.

Quiz

This is an "open book" quiz. You may refer to the text in this chapter. A good score is 8 correct. Answers are in the back of the book.

1. The sensitivity of a dynamic microphone will be reduced if it is
   (a) used with an amplifier having too much gain.
   (b) used with a monaural amplifier, rather than a stereo amplifier.
   (c) subjected to excessively high temperatures for a long time.
   (d) rated for rms power output rather than peak power output.
2. If a microphone has a cardioid pattern, the direction of least sensitivity is
   (a) sharply defined, and opposite to the direction of greatest sensitivity.
   (b) broadly defined, and at right angles to the direction of greatest sensitivity.
   (c) sharply defined, and at right angles to the direction of greatest sensitivity.
   (d) broadly defined, and opposite to the direction of greatest sensitivity.
3. A headset with a peaked audio response is best suited for use in
   (a) low-power stereo hi-fi systems.
   (b) conjunction with midrange speakers.
   (c) high-impedance applications.
   (d) two-way radio communications.
4. The lowpass-filter portion of a crossover network
   (a) prevents low-frequency audio from reaching a woofer.
   (b) prevents low-frequency audio from reaching a tweeter.
   (c) prevents high-frequency audio from reaching a woofer.
   (d) prevents high-frequency audio from reaching a tweeter.
5. Which of the following speaker parameters is of the *least* concern with regard to the quality of the sound it produces?
   (a) Efficiency.
   (b) Frequency response.
   (c) Dispersion.
   (d) Transient response.
6. Which type of speaker produces sound as a result of an alternating or fluctuating voltage between large metal plates?
   (a) Dynamic.
   (b) Piezoelectric.
   (c) Ceramic.
   (d) Electrostatic.

7. When buying a set of speakers for an amplifier system that has a particular power output rating, you should note that
   (a) rms power is generally greater than peak power.
   (b) rms power is generally less than peak power.
   (c) rms power and peak power are always the same.
   (d) rms power applies to low audio frequencies, while peak power applies to midrange and high audio frequencies.
8. Consider a whizzer with a diameter of 3/4 in. What is the highest frequency, approximately, at which good dispersion can be expected from this transducer?
   (a) 10 kHz.
   (b) 18 kHz.
   (c) 25 kHz.
   (d) 45 kHz.
9. Suppose you are told that a speaker or headset has a flat frequency response. This means that
   (a) for sine-wave signals, the sound output it produces is a little flat, that is, slightly lower in frequency than the input signal.
   (b) for sine-wave signals, the sound output it produces is in the same proportion to the audio power input over the entire range of audio frequencies.
   (c) a crossover network is required to ensure that the woofer does not receive excessive sound at bass frequencies.
   (d) it is rated for rms power handling capacity rather than peak power handling capacity over the entire range of audio frequencies.
10. Which of the following arrangements should be equipped with a crossover network to ensure optimum performance?
   (a) A dynamic speaker combination consisting of a woofer and a tweeter.
   (b) A piezoelectric transducer designed to produce or pick up ultrasound.
   (c) A crystal microphone intended for communications or computer use.
   (d) All of the above

Impedance-Matching Transformers

If you've done much audio work, you've heard the term *impedance*. Have you talked with a sales person who told you that a speaker has "low impedance" or an amplifier input has "high impedance"? In this chapter, you'll learn what impedance is. You'll also learn how audio system performance is optimized by matching the impedances of interconnected circuits and devices.

What Is Impedance?

Impedance is a quantitative expression of the opposition that a component or circuit offers to AC. It consists of two independent components: *resistance* and *reactance*. Impedance requires two numbers to be rigorously expressed.

OPPOSITION TO CURRENT

In a DC circuit, the opposition to current is simple resistance. It is expressed in ohms (symbolized Ω), and is always zero or more. The larger the value of resistance, the greater the opposition to DC.

In an AC circuit, resistance opposes the flow of current in the same way as it does for DC. But there is another phenomenon—reactance—that also opposes the flow of AC. Reactance can be *inductive*, in which case it is considered mathematically positive; or it can be *capacitive*, in which case it is considered mathematically negative. Reactance can also be equal to zero, representing the absence of inductance or capacitance in a component, circuit, device, or system.

Resistance does not normally change with the frequency of an AC wave or signal, but reactance does.

COMPLEX IMPEDANCE

Engineers use special notation to distinguish reactance from resistance. Resistance can be portrayed geometrically along a *half-line*, also called a *ray*, extending infinitely in one direction from a defined point, and corresponding to the set of all non-negative numbers. Reactance needs all the negative numbers, as well as the positive numbers and zero, to be graphically rendered. Reactance can be portrayed on a straight line extending infinitely in two opposite directions.

In order to define impedance in a pictorial way, we can place the resistance ray and the reactance line together and join them so they intersect and are perpendicular at their zero points. This forms the basis for a coordinate system known as the *complex impedance plane* or *RX plane*. (Actually it is only a half-plane!) The resistive component of impedance is symbolized R, and the reactance is symbolized jX. While R must always be a positive number or zero, X can be any number whatsoever—negative, positive, or zero. The symbol j refers to the positive square root of -1, which is a so-called *imaginary number*. It is also called the *j operator*. This rather bizarre "animal" provides engineers with a mathematical model that perfectly explains the behavior of complex impedances.

In DC circuits, resistance values R correspond to unique points on a ray. In AC circuits, impedance values, defined as $R + jX$, correspond to unique points (R,jX) on the *RX* plane (Fig. 9-1). The point where the resistance and reactance axes intersect is called the *origin*, and represents zero resistance and zero reactance. That's a perfect short circuit.

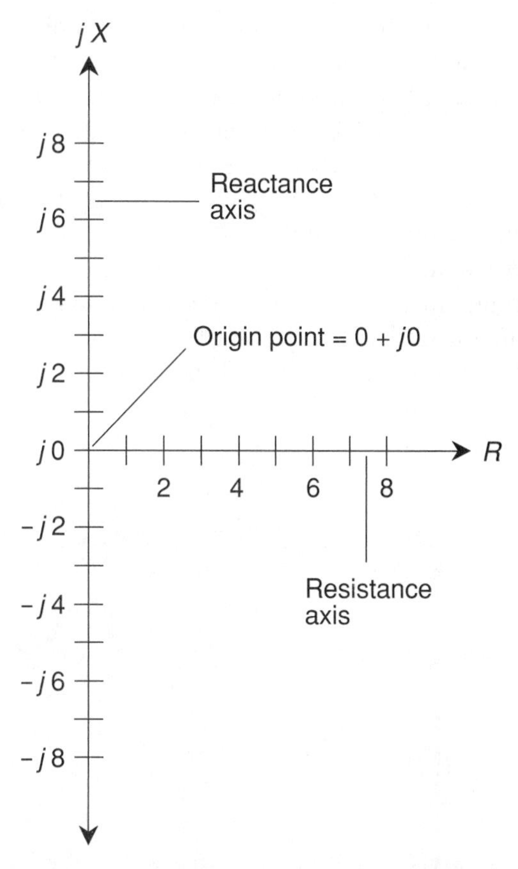

Figure 9-1 The complex impedance plane. The origin, or the point where the resistance and reactance axes intersect, represents a perfect short circuit.

ABSOLUTE-VALUE IMPEDANCE

The salespeople in electronics stores rarely talk about complex impedance. Instead, they simplify things. In most audio applications, the system reactance is zero or near zero anyhow, so it does not factor into a non-technical discussion. You will therefore be told that the "impedance" of some device or component is a certain number of ohms, and no mention will be made of its resistive or reactive components. For example, there are "8-Ω" speakers and "600-Ω" amplifier inputs. An "8-Ω" speaker actually has a complex impedance of $8 + j0$, and a "600-Ω" input circuit is designed to operate with a complex impedance of $600 + j0$.

Sometimes the italicized, uppercase letter *Z* is used in place of the expression "complex impedance" in documentation. This is what engineers mean when they say things like "Z = 8 Ω" or "Z = 8 Ω nonreactive." The expression "Z = 8 Ω" in this context, if no specific complex impedance is given, can theoretically refer to $8 + j0$, or $0 + j8$, or $0 - j8$, or any other value on a half-circle of points in the complex impedance plane that lie 8 units from the origin $0 + j0$. There are, in theory, infinitely many points on this half-circle.

Figure 9-2 shows five different *impedance vectors* as arrows that point outward from the origin and terminate on the half-circle with a radius of 8 units. All five vectors have the same length, but they point in different directions. Recall from high-school math and physics that a *vector* is a quantity having two independent properties: *magnitude* (or length) and *direction* (or orientation). Vectors provide an

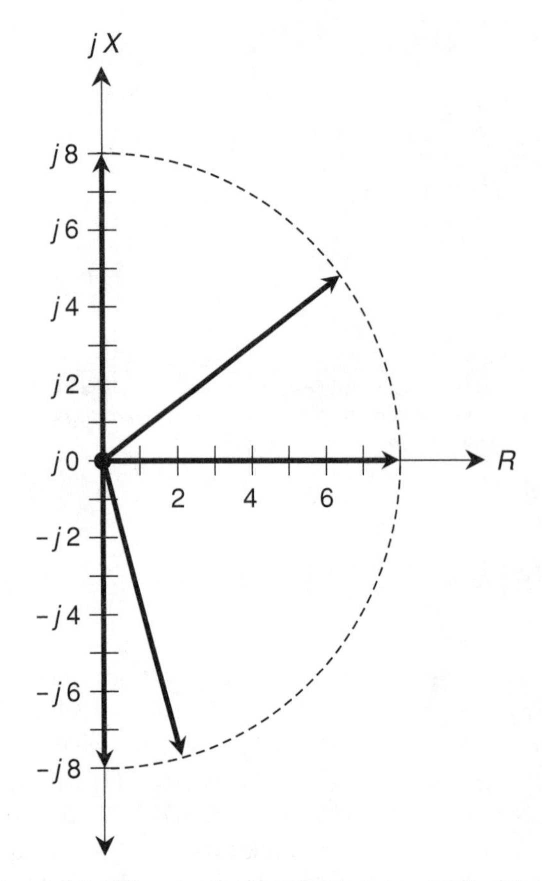

Figure 9-2 Five complex impedance vectors. Each of these vectors denotes an impedance with an absolute value of 8 Ω.

alternative way to express complex impedances. The length of a complex impedance vector is sometimes called the *absolute-value impedance*. When you imagine complex impedances as vectors having various lengths, you should be able to see that the absolute-value impedance of a component, device, or system is always at least as great as the resistive part of the complex impedance.

If you're not specifically given both the resistance and reactance values when a single-number ohmic figure is quoted for an impedance, you can assume that the engineers are talking about a *nonreactive impedance*. It is also called a *resistive impedance*. That means the impedance is a pure resistance, and that the imaginary, or reactive, component is zero. In the complex impedance plane, resistive vectors always point directly to the right (or "east") along the R axis.

Audiophiles will sometimes speak of a nonreactive impedance as "low-Z or high-Z." These are relative terms. An 8-Ω speaker would qualify as a low-Z device in most applications; a 600-Ω amplifier input would usually be considered high-Z.

PROBLEM 9-1

Suppose you have an 8-Ω speaker that exhibits no reactance over the entire range of audio frequencies. Then you connect a capacitor in series with the speaker to modify its frequency-response characteristics. Suppose this capacitor has a reactance of –4 Ω at a frequency of 1000 Hz. What is the absolute-value impedance of the speaker/capacitor combination, compared with the absolute-value impedance of the speaker alone, at 1000 Hz?

SOLUTION 9-1

The original absolute-value impedance is the resistive component of $8 + j0$, or 8 Ω. But the speaker/capacitor combination, at 1000 Hz, exhibits a complex impedance of $8 - j4$. The resulting impedance vector is longer than the original one. That means the absolute-value impedance is higher. This is shown graphically in Fig. 9-3.

Matching It

In an audio system, a *load* (such as a speaker or amplifier input) accepts all the power from a *source* (such as a microphone or amplifier output) only when the impedances of the load and source are resistive and identical. The presence of reactance, and/or the existence of a significant difference between the source and load resistances, degrades the performance. In a sound system, an *audio transformer* can ensure that the output impedance of a source is the same as the input

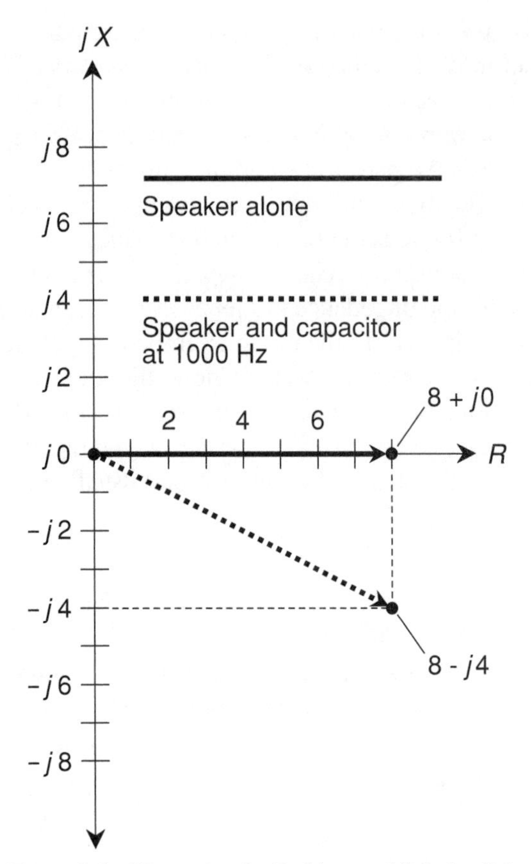

Figure 9-3 Illustration for Problem and Solution 9-1.

impedance of a load, as long as there is no reactance. This process is called *impedance matching*.

The resistive impedance a speaker or headset is usually between 4 Ω and 16 Ω. Some headsets have much higher impedances, ranging between 600 Ω and 2000 Ω. If a significant impedance mismatch exists between a speaker or headset and the output of the amplifier driving it, the amplifier efficiency suffers. In some cases, distortion can occur, even at low volume levels, when the impedances are not properly matched. This phenomenon can be expected, for example, if a high-impedance headset is used with a system designed for low-impedance headsets.

In audio systems, specially designed *broadband transformers* are the preferred method of impedance matching. Transformers, unlike inductance-capacitance (*LC*) networks, can function uniformly over a wide range of frequencies. In radio trans-

mitters and antennas where the frequency is constant, *LC* networks are often used for impedance matching, but that scheme is not of much use in a hi-fi system where the frequency range equals or exceeds three orders of magnitude—from 20 Hz or less to 20 kHz or more.

Transformer Basics

When two wires are near each other and one of them carries an alternating or fluctuating current, AC is induced in the other wire. This effect is known as *electromagnetic induction*. All *AC transformers* work according to this principle. If the first wire carries sine-wave AC of a certain frequency, then the *induced current* is sine-wave AC of the same frequency in the second wire.

COILS AND FORMS

If two lengths of wire are wound into coils and placed along a common axis (Fig. 9-4), the *magnetic coupling* is better than is the case with straight, parallel wires, because electromagnetic induction takes place to a greater extent. The alternating

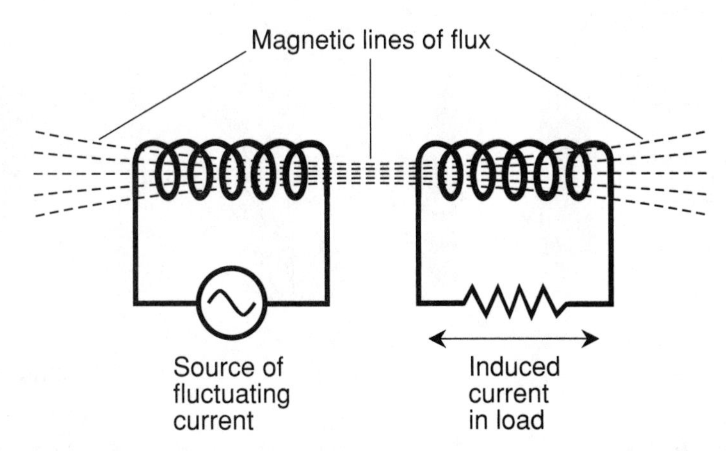

Figure 9-4 Magnetic lines of flux between two aligned coils of wire. Alternating or fluctuating current in the left-hand coil produces AC in the right-hand coil, and therefore also in the load.

magnetic lines of flux from the first coil produce AC of the same frequency, and having the same wave shape, in the second coil. Optimum *inductive coupling,* or efficiency of induced-current transfer, is obtained when two lengths of insulated wire are wound on a single *coil form,* with one winding placed directly over the other.

In a typical audio transformer, the coil form is made of laminated or powdered iron, which concentrates the magnetic lines of flux and reduces the number of coil turns necessary for effective operation. In addition to this, the form itself is typically bent into a shape called an *E core,* so the magnetic flux doesn't stray very much from the immediate vicinity of the coils. This keeps the transformer from inducing currents in, or being affected by currents from, the surrounding components and wiring.

Most broadband audio transformers have E cores with primary and secondary windings wound around the cross bars, as shown in Fig. 9-5. At A in the figure, a disassembled E core is shown. When both wires are wound around the central horizontal cross bar (Fig. 9-5B), the transformer is said to have a *shell winding.* Reduced *capacitive coupling* (an unwanted phenomenon, as opposed to inductive coupling) between the windings can be obtained by putting the coils on opposite horizontal cross bars (Fig. 9-5C). This is called a *core winding.* Core winding works better than shell winding for broadband audio transformers intended for impedance matching.

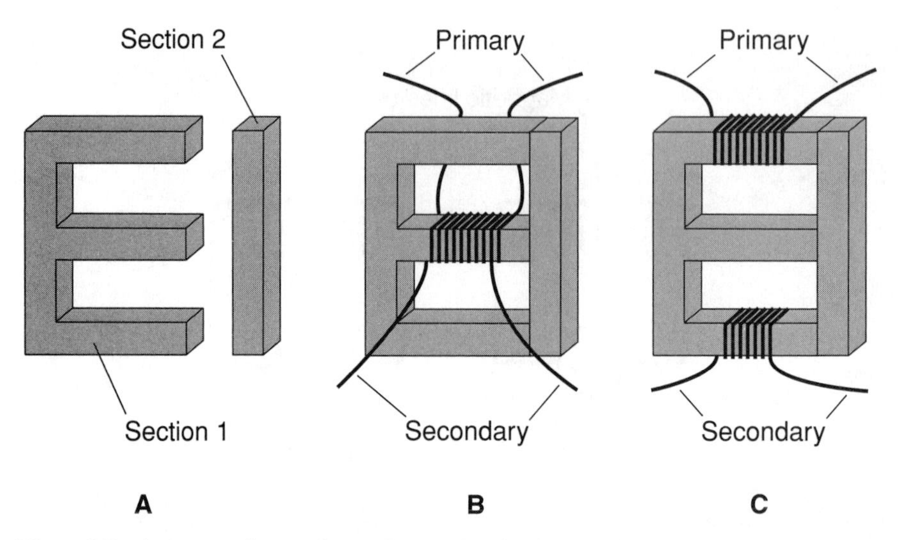

Figure 9-5 At A, an audio transformer E core, showing both sections before they are clamped together. At B, the shell winding configuration. At C, the core winding configuration.

PRIMARY AND SECONDARY

In a transformer, the first coil, to which the input AC voltage or signal is applied, is called the *primary winding*. The second coil, from which the output voltage or signal is taken, is known as the *secondary winding*. These are often spoken of as the *primary* and the *secondary*. In a *step-down* transformer where the secondary has fewer turns than the primary, the voltage across the secondary is less than the voltage across the primary. In a *step-up* transformer where the secondary has more turns than the primary, the voltage across the secondary is greater than the voltage across the primary.

The windings of a transformer have inductance because they are coils. The optimum inductances of the primary and secondary for a particular audio transformer depend on the frequency of operation, and also on the resistive part of the impedance in the circuit. As the frequency increases, the optimum inductance decreases, if all other circuit factors are held constant. At high resistive impedances, more inductance is generally needed than at low resistive impedances. The best transformer for a particular audio system must usually be found by trial and error during the design phase.

TURNS RATIO AND VOLTAGE

The primary voltage in a transformer is symbolized E_{pri}, and the secondary voltage is symbolized E_{sec}. Unless otherwise stated, rms voltages are always specified. The *primary-to-secondary turns ratio* in a transformer is the ratio of the number of turns in the primary, T_{pri}, to the number of turns in the secondary, T_{sec}. This ratio is written $T_{pri}:T_{sec}$ or T_{pri}/T_{sec}.

In a transformer with excellent primary-to-secondary inductive coupling, the following relationship holds:

$$E_{pri}/E_{sec} = T_{pri}/T_{sec}$$

That is, the *primary-to-secondary voltage ratio* is always equal to the primary-to-secondary turns ratio (Fig. 9-6). In a step-down transformer, T_{pri}/T_{sec} and E_{pri}/E_{sec} are theoretically equal, and the ratio is greater than 1. In a step-up transformer, T_{pri}/T_{sec} and E_{pri}/E_{sec} are theoretically equal, and the ratio is smaller than 1.

Sometimes the *secondary-to-primary turns ratio* is specified. This is written T_{sec}/T_{pri}. When this parameter is given, the numerators and denominators are inverted compared with the primary-to-secondary turns and voltage ratios. Thus, in a step-down transformer, T_{sec}/T_{pri} and E_{sec}/E_{pri} are theoretically equal, and the ratio is smaller than 1. In a step-up transformer, T_{sec}/T_{pri} and E_{sec}/E_{pri} are theoretically equal, and the ratio is greater than 1.

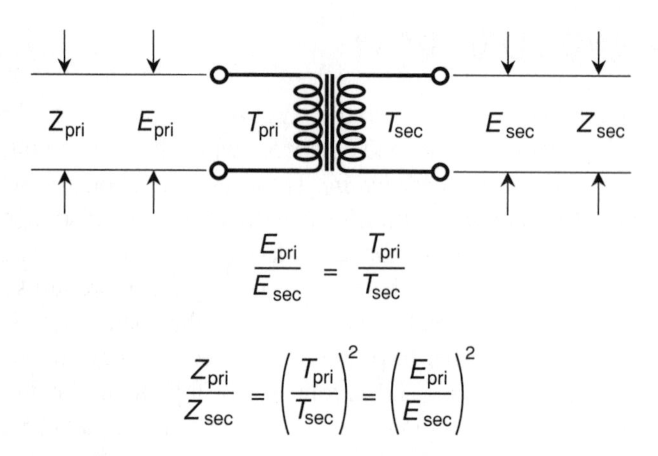

Figure 9-6 In an audio transformer, the primary voltage (E_{pri}) and secondary voltage (E_{sec}), as well as the primary impedance (Z_{pri}) and secondary impedance (Z_{sec}), depend on the number of turns in the primary winding (T_{pri}) and the number of turns in the secondary winding (T_{sec}).

PROBLEM 9-2

Suppose a transformer has a primary-to-secondary turns ratio of exactly 9:1. The AC voltage at the primary is 90 V rms. Is this a step-up transformer or a step-down transformer? What is the voltage across the secondary?

SOLUTION 9-2

This is a step-down transformer, because T_{pri}/T_{sec} is greater than 1. To determine the voltage across the secondary, plug in the numbers in the above equation and solve for E_{sec}, as follows:

$$\begin{aligned} E_{pri}/E_{sec} &= T_{pri}/T_{sec} \\ 90/E_{sec} &= 9/1 \\ 1/E_{sec} &= 9/90 \\ E_{sec} &= 90/9 \\ &= 10 \text{ V rms} \end{aligned}$$

PROBLEM 9-3

Consider a transformer with a primary-to-secondary turns ratio of exactly 1:9. The voltage at the primary is 0.9 V rms. Is this a step-up transformer or a step-down transformer? What is the voltage at the secondary?

SOLUTION 9-3

This is a step-up transformer, because T_{pri}/T_{sec} is less than 1. Plug in numbers and solve for E_{sec}, as follows:

$$0.9/E_{sec} = 1/9$$
$$E_{sec}/0.9 = 9$$
$$E_{sec} = 9 \times 0.9$$
$$= 8.1 \text{ V rms}$$

Impedance-Transfer Ratio

The *impedance-transfer ratio* of a transformer varies according to the square of the turns ratio, and also according to the square of the voltage-transfer ratio (see Fig. 9-6 again). If the primary (source) and secondary (load) impedances are purely resistive and are denoted Z_{pri} and Z_{sec}, then the following relations hold:

$$Z_{pri}/Z_{sec} = (T_{pri}/T_{sec})^2$$
$$Z_{pri}/Z_{sec} = (E_{pri}/E_{sec})^2$$

The inverses of these formulas, in which the turns ratio or voltage-transfer ratio are expressed in terms of the impedance-transfer ratio, are:

$$T_{pri}/T_{sec} = (Z_{pri}/Z_{sec})^{1/2}$$
$$E_{pri}/E_{sec} = (Z_{pri}/Z_{sec})^{1/2}$$

where the 1/2 power represents the positive square root.

PROBLEM 9-4

Suppose you have an audio amplifier whose output is 800 Ω, and you want to use a headset that has an impedance of 8 Ω. What is the turns ratio of a transformer that can match the impedances in this situation?

SOLUTION 9-4

You will need a transformer with a turns ratio of 10:1. The square of the turns ratio is 100:1, which is the equivalent of 800:8. The transformer primary is connected to the 800-Ω audio amplifier output, and the transformer secondary is connected to the 8-Ω headset. Again, it is important to note that this works only when there is no reactance in the system.

PROBLEM 9-5

Consider a situation in which a transformer is needed to match an input impedance of 50 Ω, purely resistive, to an output impedance of 300 Ω, also purely resistive. What is the required turns ratio T_{pri}/T_{sec}?

SOLUTION 9-5

The required transformer will have a step-up impedance ratio of $Z_{pri}/Z_{sec} = 50/300 = 1/6$. From the above formulas:

$$\begin{aligned} T_{pri}/T_{sec} &= (Z_{pri}/Z_{sec})^{1/2} \\ &= (1/6)^{1/2} \\ &= (1^{1/2})/(6^{1/2}) \\ &= 1/2.45 \end{aligned}$$

PROBLEM 9-6

Suppose a transformer has a primary-to-secondary turns ratio of exactly 4:1. The load, connected to the transformer output, is a pure resistance of 37.5 Ω. What is the impedance at the primary?

SOLUTION 9-6

The impedance-transfer ratio is equal to the square of the turns ratio. Therefore:

$$\begin{aligned} Z_{pri}/Z_{sec} &= (T_{pri}/T_{sec})^2 \\ &= (4/1)^2 \\ &= 4^2 \\ &=16 \end{aligned}$$

We know that the secondary impedance, Z sec is 37.5 Ω. Thus:

$$\begin{aligned} Z_{pri} &= 16.0 \times Z_{sec} \\ &= 16.0 \times 37.5 \\ &= 600\ \Omega \end{aligned}$$

Quiz

This is an "open book" quiz. You may refer to the text in this chapter. A good score is 8 correct. Answers are in the back of the book.

1. Impedance matching requires that the reactances in the input and output
   (a) be identical.
   (b) be equal and opposite.
   (c) both be equal to zero.
   (d) exist in the same ratio as the resistances.
2. In Fig. 9-7, which component could represent a typical 8-Ω speaker operating normally at AF?
   (a) Component *X*.
   (b) Component *Y*.
   (c) Component *Z*.
   (d) None of them

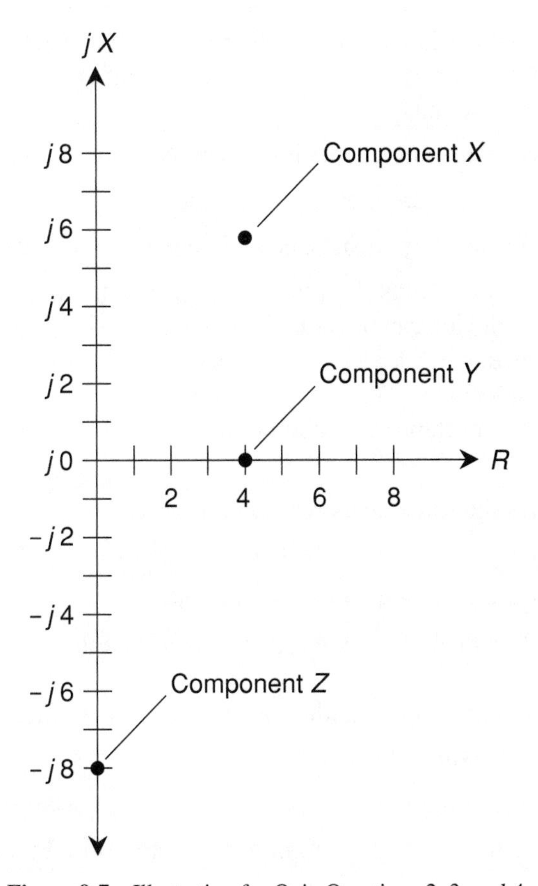

Figure 9-7 Illustration for Quiz Questions 2, 3, and 4.

3. In Fig. 9-7, which component exhibits a purely resistive impedance?
   (a) Component *X*.
   (b) Component *Y*.
   (c) Component *Z*.
   (d) None of them

4. In Fig. 9-7, which component exhibits capacitive reactance?
   (a) Component *X*.
   (b) Component *Y*.
   (c) Component *Z*.
   (d) None of them

5. Suppose you are told that the absolute-value impedance of a certain device is less than the resistive component of the complex impedance. From this, you can conclude that
   (a) the device has pure capacitive reactance.
   (b) the device has pure inductive reactance.
   (c) the device has high resistance and negative reactance.
   (d) your informant must be mistaken, because the absolute-value impedance of a device can never be less than the resistive part of the complex impedance.

6. When there is reactance in a component, the resistance of that component, in ohms,
   (a) is always the same as its reactance in ohms.
   (b) is always greater than its reactance in ohms.
   (c) is always less than its reactance in ohms.
   (d) can be less than, the same as, or greater than its reactance in ohms.

7. Capacitive coupling in an audio transformer can be minimized by
   (a) the shell winding method.
   (b) the core winding method.
   (c) using an air-core coil form, rather than an iron-core form.
   (d) winding the secondary right on top of the primary.

8. What is the primary-to-secondary impedance-transfer ratio of the broadband audio transformer shown in Fig. 9-8, assuming it is operated in the AF range?
   (a) 4:1.
   (b) 2:1.
   (c) 1:2.
   (d) 1:4.

9. If a 500-Hz sine-wave signal having 16 V rms is applied to the primary winding of the broadband audio transformer shown in Fig. 9-8, what will be the rms voltage of the audio output signal?
   (a) 64 V rms.
   (b) 32 V rms.
   (c) 8 V rms.
   (d) 4 V rms.

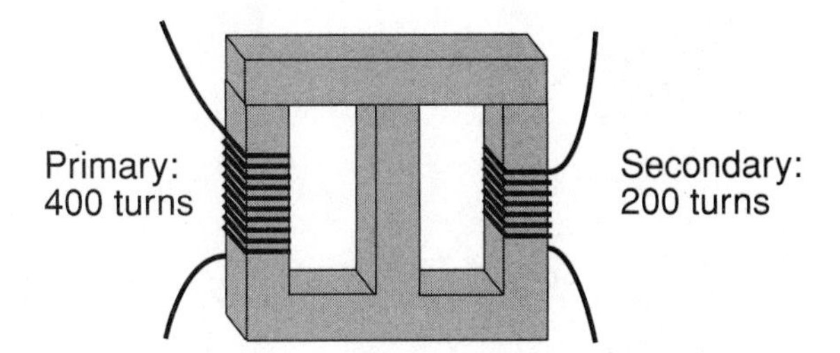

Figure 9-8 Illustration for Quiz Questions 8 and 9.

10. Suppose you want to use an impedance-matching transformer to operate a 4-Ω microphone with a preamplifier rated at 900 Ω input impedance. What should the primary-to-secondary turns ratio of the transformer be, assuming you connect the microphone to the primary and the amplifier input to the secondary?
   (a) 1/15.
   (b) 1/225.
   (c) 15/1.
   (d) 225/1.

CHAPTER 10

Transistorized Amplifiers and Oscillators

Now that you've learned a little about transistors, you're ready to learn about amplifiers and oscillators that use them. But before we get into that, let's look a little more closely at the decibel, which was briefly introduced earlier in this book.

The Decibel Revisited

Electronic circuits, like human ears, often react to varying signals according to the logarithm of the actual signal strength. The decibel is used to define relative intensity for all sorts of effects and phenomena, even radiant energy such as visible light and radio waves! Here, we'll examine how it is used to compare relative AC signal voltage, current, and power in electronic circuits.

GAIN AND LOSS

In the decibel scheme, increases in amplitude are assigned *positive gain* values, and decreases in amplitude are assigned *negative gain* values. A negative gain figure is sometimes expressed as a *loss* by removing the minus sign. If the output signal amplitude from a circuit is +6 dB relative to the input signal amplitude, then the output is stronger than the input. If the output is at –14 dB relative to the input, then the output is weaker than the input. In the first case, the circuit has a gain of +6 dB. In the second case, we can say that the circuit has a gain of –14 dB or a loss of 14 dB.

The decibel has meaning only when two or more signals are compared. For sound, a volume increase or decrease of plus or minus one decibel (±1 dB) is roughly the smallest change a listener can detect if the change is expected. If the change is not expected, the smallest difference a listener can notice is about ±3 dB.

VOLTAGE GAIN

Consider a circuit with an rms AC input voltage of E_{in} and an rms AC output voltage of E_{out}, specified in the same units as E_{in}. Then the *voltage gain* of the circuit, in decibels, is given by this formula:

$$\text{Gain (dB)} = 20 \log (E_{out}/E_{in})$$

PROBLEM 10-1

Suppose a circuit has an rms AC input of 1.00 V and an rms AC output of 14.0 V. What is the gain in decibels? Round off the answer to three significant digits.

SOLUTION 10-1

First, find the ratio E_{out}/E_{in}. Because E_{out} = 14.0 V rms and E_{in} = 1.00 V rms, the ratio is 14.0/1.00, or 14.0. Next, find the logarithm of 14.0. This is about 1.146. . . . (The three dots indicate extra digits introduced by the calculator. You can leave them in until the final roundoff.) Finally, multiply this number by 20, getting something like 22.922. . . . When we round this off to three significant digits, we get 22.9 dB.

PROBLEM 10-2

Suppose a circuit has an rms AC input voltage of 24.2 V and an rms AC output voltage of 19.9 V. What is the gain in decibels? Round off the answer to three significant digits.

SOLUTION 10-2

Find the ratio E_{out}/E_{in} = 19.9/24.2 = 0.822. . . . Find the logarithm of this. You should get a value of −0.0849. . . . The gain is 20 times that, or −1.699. . . dB, which rounds off to −1.70 dB.

CURRENT GAIN

Current gain is calculated in the same way as voltage gain. If I_{in} is the rms AC input current and I_{out} is the rms AC output current specified in the same units as I_{in}, then:

$$\text{Gain (dB)} = 20 \log (I_{out}/I_{in})$$

POWER GAIN

The *power gain* through a circuit, in decibels, is calculated according to a slightly different formula. If P_{in} is the input signal power and P_{out} is the output signal power expressed in the same units as P_{in}, then:

$$\text{Gain (dB)} = 10 \log (P_{out}/P_{in})$$

The *coefficient* (that is, the factor by which the logarithm is to be multiplied) in the formula for power gain is 10, whereas for voltage and current gain it is 20.

PROBLEM 10-3

Suppose an audio power amplifier has an input of 5.03 W and an output of 125 W. What is the gain in decibels? Round off the answer to two significant digits.

SOLUTION 10-3

Find the ratio P_{out}/P_{in} = 125/5.03 = 24.85. . . . Then find the logarithm of this. You should get 1.395. . . . Then multiply by 10 to get 13.95. . . . This rounds off to 14 dB.

PROBLEM 10-4

Suppose an *attenuator* (a circuit designed deliberately to produce power loss) provides 10 dB power reduction. The input power is 94 W. What is the output power? Round the answer off to two significant digits.

SOLUTION 10-4

An attenuation of 10 dB represents a gain of −10 dB. We know that P_{in} = 94 W, so the unknown in the power gain formula is P_{out}. We must solve for P_{out} in this formula:

$$-10 = 10 \log (P_{out}/94)$$

First, divide each side by 10, getting:

$$-1 = \log (P_{out}/94)$$

To solve this, we must take the *base-10 antilogarithm*, also known as the *antilog*, or the *inverse log*, of each side. The antilog function "undoes" the log function. The antilog of a value x is written antilog x. It can also be denoted as $\log^{-1} x$ or 10^x. Antilogarithms can be determined with any good scientific calculator. The solution process goes like this:

$$\text{antilog} (-1) = \text{antilog} [\log (P_{out}/94)]$$
$$0.1 = P_{out}/94$$
$$94 \times 0.1 = P_{out}$$
$$P_{out} = 9.4 \text{ W}$$

DECIBELS AND IMPEDANCE

When determining the voltage gain (or loss) and the current gain (or loss) for a circuit in decibels, you should expect to get the same figure for both parameters only when the input impedance is identical to the output impedance. If the input and output impedances differ, the voltage gain or loss is generally not the same as the current gain or loss.

Consider how transformers work. A step-up transformer, in theory, has voltage gain, but this alone doesn't make a signal more powerful. A step-down transformer can exhibit theoretical current gain, but again, this alone does not make a signal more powerful. In order to make a signal more *power*ful, a circuit must increase the signal *power*—the *product* of the voltage and the current!

When determining power gain (or loss) for a particular circuit in decibels, the input and output impedances don't matter. In this sense, positive power gain always represents a real-world increase in signal strength. Similarly, negative power gain (or power loss) always represents a true decrease in signal strength.

Basic Amplifier Circuits

In audio applications, amplifiers must produce output waveforms that are faithful (although magnified) reproductions of the input waveforms. Here are the types of

transistorized amplifiers most often found in audio systems. Some circuits are more efficient than the ones described in this section, but they invariably produce *waveform distortion*. That's sometimes tolerable in radio transmitters, but not in audio systems!

A SIMPLE BIPOLAR TRANSISTOR AMPLIFIER

Figure 10-1 is a schematic diagram of a *common emitter amplifier* that uses an NPN bipolar transistor. The input signal passes through C_2 to the base. Resistors R_2 and R_3 provide the base bias. Resistor R_1 and capacitor C_1 allow for the emitter to maintain a DC voltage relative to ground, while keeping it grounded for audio signals. Resistor R_1 also limits the current through the transistor. The AC component of the audio output signal goes through capacitor C_3. Resistor R_4 keeps the audio output signal from being short-circuited through the power supply.

At audio frequencies in this amplifier, capacitor values can range from a few microfarads up to about 100 μF. The optimum resistor values depend on the application. In the case of a weak-signal circuit such as a high-impedance microphone preamplifier, typical values are 220 Ω for R_1, 2.2 kΩ for R_2, 4.7 kΩ for R_3, and 820 Ω for R_4. For moderate-level power amplification, these values are reduced by a factor of several times. If high audio output power is required, a better choice of circuit is the *push-pull amplifier* described on pages 170 and 171.

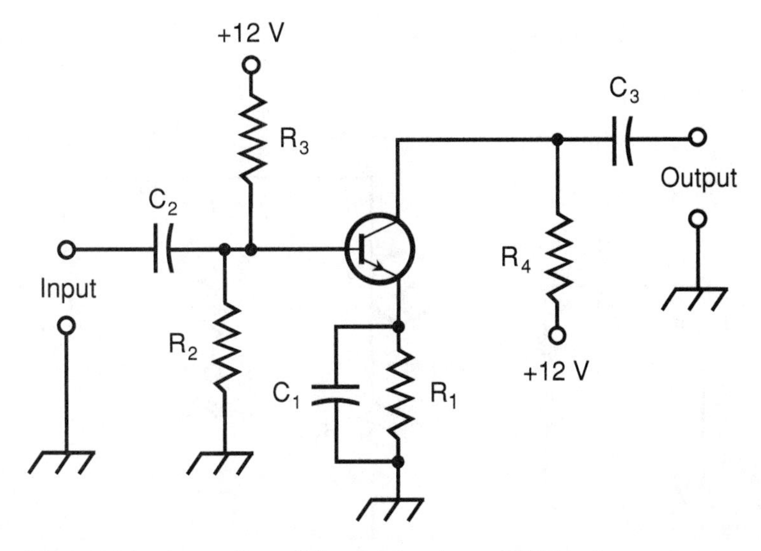

Figure 10-1 A generic amplifier circuit using an NPN bipolar transistor. Component designators are discussed in the text.

A SIMPLE JFET AMPLIFIER

Figure 10-2 shows an N-channel JFET hooked up as a *common source amplifier*. The input signal passes through C_2 to the gate. Resistor R_2 provides the bias. Resistor R_1 and capacitor C_1 give the source a DC voltage relative to ground, while grounding it for signals. The AC component of the audio output signal goes through C_3. Resistor R_3 keeps the audio output signal from being short-circuited through the power supply.

A JFET has a high input impedance, and therefore the value of C_2 should usually be small. If the device is a MOSFET, the input impedance is higher still, and C_2 will be smaller yet, sometimes 1 pF or less. The resistor values depend on the application. In some instances, R_1 and C_1 are not used, and the source is grounded directly. If R_1 is used, its optimum value depends on the input impedance and the bias desired. For weak-signal amplification (an ideal application for a JFET), typical values are 220 Ω for R_1, 2.2 kΩ for R_2, and 820 Ω for R_3.

THE CLASS-A AMPLIFIER

With the previously mentioned component values, the amplifier circuits in Figs. 10-1 and 10-2 operate in a mode known as *class A*. An amplifier in this mode is linear. As you learned in Chapter 7, that means the output waveform has the same shape as (although a much greater amplitude than) the input waveform.

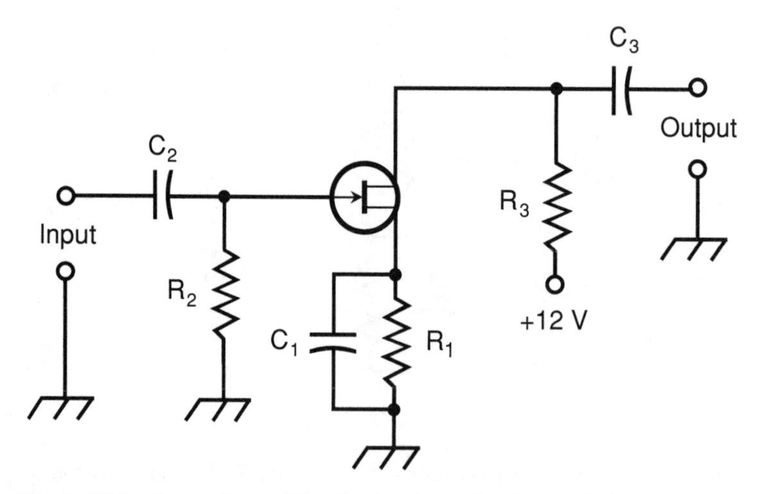

Figure 10-2 A generic amplifier circuit using an N-channel JFET. Component designators are discussed in the text.

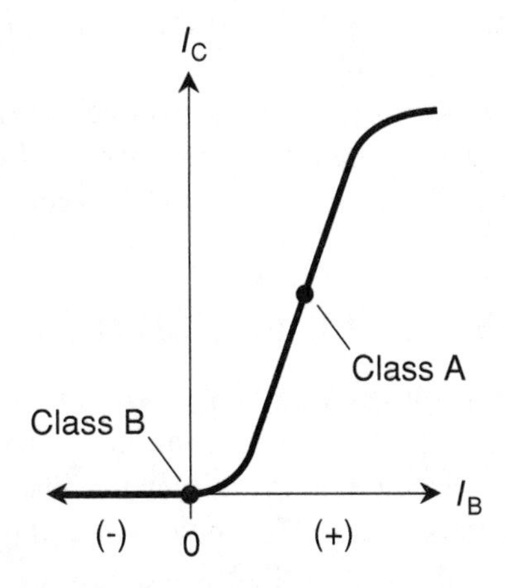

Figure 10-3 Operating DC bias points for class A and class B using an NPN bipolar transistor.

For class-A operation with a bipolar transistor, the bias must be such that, with no signal input, the device is near the middle of the straight-line portion of the I_C vs. I_B (collector current vs. base current) curve. Therefore, collector current flows even if there is no input signal. This is shown graphically for an NPN transistor in Fig. 10-3. For PNP devices, reverse the polarity signs. With a JFET or MOSFET, the bias must be such that, with no signal input, the device is near the middle of the straight-line part of the I_D vs. E_G (drain current vs. gate voltage) curve. That means drain current flows even if there is no input signal. The class-A arrangement for an N-channel JFET is shown in Fig. 10-4. For P-channel devices, reverse the polarity signs.

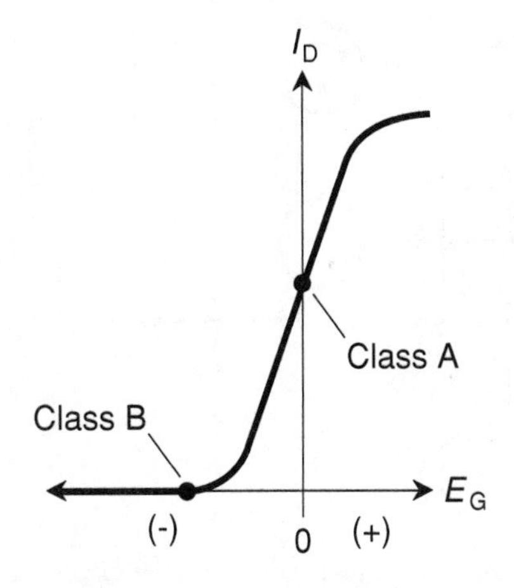

Figure 10-4 Operating DC bias points for class A and class B using an N-channel JFET.

In a class-A amplifier, it is important that the input signal not be too strong. An excessively strong input signal will drive the device out of the straight-line part of the characteristic curve during part of the signal cycle. When this occurs, the output waveform will no longer be a faithful reproduction of the input waveform, and the amplifier will become nonlinear. The result: distortion!

THE CLASS-B PUSH-PULL AMPLIFIER

When a bipolar transistor is biased exactly at cutoff, or an FET is biased exactly at pinchoff under zero-input-signal conditions, an amplifier is working in *class B*. These operating points are labeled on the curves in Figs. 10-3 and 10-4.

A class-B amplifier offers better efficiency than a class-A amplifier using the same type of device, and at the same voltages and power levels. Unfortunately, this mode of amplification produces waveform distortion. Any *single-ended* class-B amplifier (that is, a circuit that uses only one transistor) will exhibit nonlinearity with respect to the input signal wave. In its basic form, therefore, this type of amplifier is no good for hi-fi audio.

The nonlinearity problem with class-B amplification can be overcome if two transistors are connected in a so-called *push-pull configuration*. An example of a push-pull audio power amplifier using two NPN bipolar transistors is illustrated in Fig. 10-5. Resistor R_1 limits the current through the transistors. Capacitor C_1 keeps

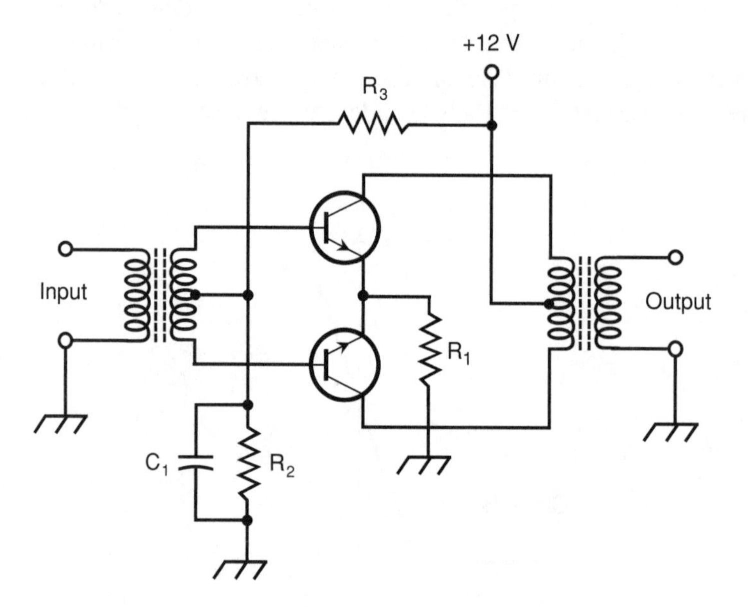

Figure 10-5 A class-B push-pull audio amplifier circuit using two NPN bipolar transistors. Component designators are discussed in the text.

the input transformer center tap at signal ground, while allowing for adjustment of the DC base bias. Resistors R_2 and R_3 have values such that they bias the transistors precisely at their cutoff points.

In a push-pull circuit, the two transistors must have characteristics that are as nearly identical as possible. Their part numbers must be the same, and ideally they should be chosen *by experiment* to ensure that their characteristic curves are closely matched.

The class-B push-pull configuration combines the efficiency of class-B operation with the low distortion of class-A operation. Its main disadvantage is that it needs two center-tapped transformers, one at the input and the other at the output. This makes push-pull amplifiers somewhat more bulky, massive, and expensive than class-A amplifiers with the same power output.

Output, Input, and Efficiency

The power ratings and efficiency of an audio amplifier can be defined according to various specifications. Four of the most basic parameters are described here.

AUDIO POWER OUTPUT

The *audio power output* of an amplifier must be measured by means of a specialized *audio-frequency* (AF) *wattmeter*. When there is no signal input to an amplifier, there is no signal output, and therefore the power output is zero. This is true no matter what the class of amplification. The greater the signal input, in general, the greater the useful audio output of a power amplifier, up to a certain point. Audio power output is measured in watts. For very low-power circuits, it can be in milliwatts; for high-power circuits it is sometimes given in kilowatts.

AUDIO POWER INPUT

The *audio power input* of an amplifier, like the audio power output, must be measured with an AF wattmeter. Under ideal conditions, the audio power input to an amplifier is the same as the audio power output from the preceding stage. But in the real world, there is always some loss in the coupling circuits, particularly transformers. For this reason, the audio power input to a given stage is always a little less than the audio power output from the preceding stage.

DC POWER INPUT

Imagine that you connect an ammeter or milliammeter in series with the collector or drain of an amplifier and the power supply. While the amplifier is in operation, this meter will have a certain reading. The reading might appear constant, or it might fluctuate with changes in the input signal level. The *DC collector power input* to a bipolar-transistor amplifier circuit is the product of the collector current (I_C) and the collector voltage (E_C). Similarly, for an FET, the *DC drain power input* is the product of the drain current (I_D) and the drain voltage (E_D). These power figures can be further categorized as *average* or *peak* values. This discussion involves only average power.

The DC collector or drain power input can be high even when there is no audio input signal. A class-A circuit operates this way. In fact, when a signal is applied to a class-A amplifier, the meter reading, and therefore the DC collector or drain power input, does not change compared to the value under conditions of no input signal. In class-B operation, there is no current, and therefore zero DC collector or drain power input, when there is no input signal. The current and the DC power input increase with increasing signal input.

The DC collector or drain power input to a power amplifier is usually measured in watts, the product of amperes and volts. It can be indicated in milliwatts for low-power amplifiers, or kilowatts for high-power amplifiers.

EFFICIENCY

The *efficiency* of a power amplifier is the ratio of the audio power output to the DC collector or drain power input.

In a bipolar-transistor amplifier, suppose that P_C is the DC collector power input, and P_{out} is the audio power output in the same units as P_C. For an FET amplifier, let P_D be the DC drain power input, and let P_{out} be the audio power output in the same units as P_D. Then the efficiency (eff) of the bipolar transistor amplifier is given by:

$$\text{eff} = P_{out}/P_C$$

For the FET circuit, the efficiency is:

$$\text{eff} = P_{out}/P_D$$

These quotients are always between 0 and 1. Efficiency is often expressed as a percentage instead of a ratio, so the above formulas are modified as follows:

$$\text{eff}_{\%} = 100\ P_{\text{out}}/P_{\text{C}}$$

and

$$\text{eff}_{\%} = 100\ P_{\text{out}}/P_{\text{D}}$$

Class-A amplifiers have efficiency figures from 25 to 40 percent, depending on the nature of the input signal and the type of transistor used. Class-B amplifiers are typically 50 to 65 percent efficient.

PROBLEM 10-5

Suppose a bipolar-transistor amplifier has a DC collector input of 50 W and an audio power output of 20 W. What is the efficiency in percent?

SOLUTION 10-5

Use the formula for the efficiency of a bipolar transistor amplifier expressed as a percentage:

$$\begin{aligned} \text{eff}_{\%} &= 100P_{\text{out}}/P_{\text{C}} \\ &= 100 \times 20/50 \\ &= 100 \times 0.4 \\ &= 40\% \end{aligned}$$

PROBLEM 10-6

Suppose an FET amplifier is 60 percent efficient. If the audio power output is 3.5 W, what is the DC drain power input? Round the answer off to the nearest tenth of a watt.

SOLUTION 10-6

Plug in values to the formula for the efficiency of an FET amplifier expressed as a percentage. The resulting equation is solved as follows:

$$\begin{aligned} 60 &= 100 \times 3.5/P_{\text{D}} \\ 60 &= 350/P_{\text{D}} \\ 60/350 &= 1/P_{\text{D}} \\ P_{\text{D}} &= 350/60 \\ &= 5.8\ \text{W} \end{aligned}$$

Other Amplifier Considerations

Class-A amplifiers do not, in theory, draw any power from the signal source to produce significant output power. This is one of the advantages of this class of operation. It is only necessary that a certain voltage be present at the control electrode (the base, gate, emitter, or source) for these circuits to produce useful output signal power. Class-B amplifiers do require that the source deliver some power. This power is called *drive* or *driving power*.

OVERDRIVE

In an audio amplifier where fidelity is of concern, it is important that the driving signal not be too strong. An excessive amount of input voltage or power is called *overdrive*. When an amplifier is overdriven, distortion occurs in the output signal.

A sine-wave generator and an oscilloscope can be used to determine whether or not an amplifier is being overdriven. The sine-wave generator is connected to the amplifier input, and the vertical input of the scope is connected to the amplifier output. The waveform of the output signal is examined. The output waveform for a particular class of amplifier always has a characteristic shape. Overdrive is indicated by a form of distortion known as *flat topping*.

In Fig. 10-6A, the output signal waveshape for a properly operating class-A or class-B push-pull audio amplifier is shown, when a pure sine wave is applied to the input. In Fig. 10-6B, the output of the same amplifier is shown under conditions of excessive drive, again with a pure sine wave. Note that the wave peaks are blunted.

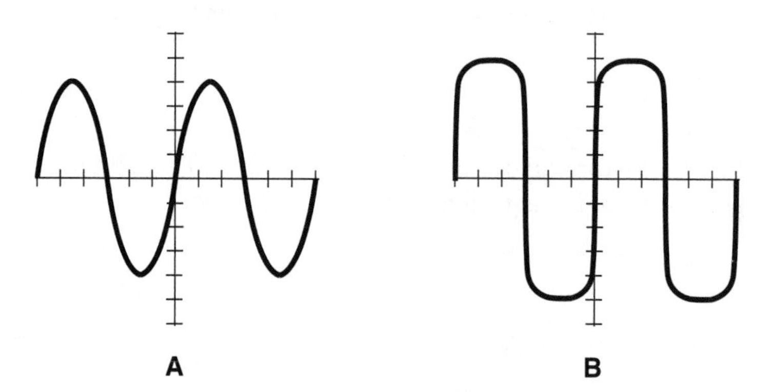

Figure 10-6 At A, the output of an amplifier with an input sine wave of the proper amplitude. At B, the output of the same amplifier with flat topping produced by excessive drive.

This causes poor fidelity and exaggerated output at harmonics of the audio input frequencies. Excessively strong harmonic signals can damage some speakers, and they compound the distortion problem because they mix with each other and with the intended output waves, producing "false waves" at frequencies not in the original sound source. The efficiency of the amplifier can be degraded, as well.

FREQUENCY RESPONSE

High-fidelity audio amplifiers, of the kind used in music systems, should be capable of producing constant gain at all frequencies within the range of human hearing from 20 Hz to 20 kHz. This is a frequency span of 1000:1 or three *orders of magnitude* (powers of 10). The best high-end audio amplifiers are capable of constant gain from *infrasonic frequencies* (less than 20 Hz) to *ultrasonic frequencies* (more than 20 kHz). Contrast this with audio amplifiers for voice communication systems such as two-way radios and telephones, where the gain must be constant only within a frequency range from 300 Hz to 3 kHz, which represents only one order of magnitude.

TRANSFORMER COUPLING

Transformers can be used to transfer (or *couple*) signals from one stage to the next in a cascaded amplifier system (also known as an *amplifier chain*). An example of *transformer coupling* is shown in Fig. 10-7. Capacitors C_1 and C_2 keep one end

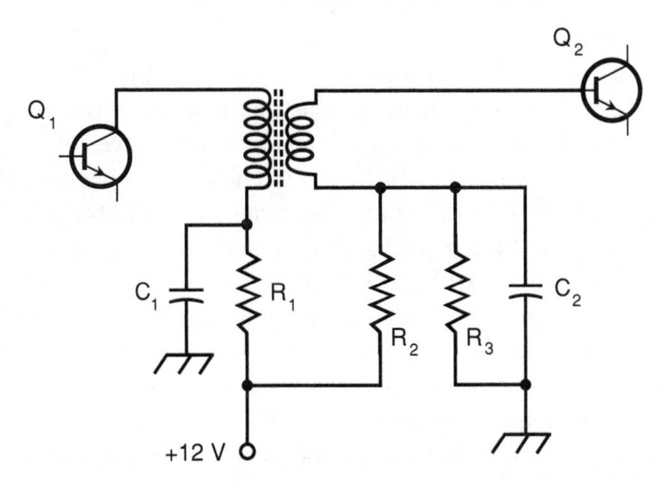

Figure 10-7 An example of transformer coupling between two audio amplifier stages. Component designators are discussed in the text.

of the transformer primary and secondary at signal ground while allowing for some DC voltage for the transistor electrodes. Resistor R_1 limits the current through the first transistor, Q_1. Resistors R_2 and R_3 provide the proper base bias for transistor Q_2. When the transformer has the correct turns ratio, the output impedance of Q_1 is perfectly matched to the input impedance of Q_2.

WEAK-SIGNAL APPLICATIONS

Some amplifiers are designed specifically for weak-signal work. A good example is a microphone preamplifier. Another example is an amplifier designed for use with a *pickup transducer* in a sonar system. In weak-signal scientific applications, amplifier sensitivity is as important—in some cases more important—than linearity.

Amplifier sensitivity is determined by the gain and also by the *noise figure*, a measure of how well a circuit can amplify desired signals while generating a minimum of electronic noise. The concept of noise figure is discussed in Chapter 12.

All bipolar transistors or FETs generate some *white noise* because of the movement of the charge carriers among the atoms. In general, JFETs produce less of this internal noise than bipolar transistors. Gallium arsenide FETs, also called *GaAsFETs* (pronounced "gasfets"), are the least noisy of all.

ATTENUATORS

An *attenuator* is a component, circuit, or device that reduces the strength of a signal. Attenuators are sometimes used in the input circuits of audio power amplifiers to prevent overdrive.

Attenuators for hi-fi use are designed to operate over the entire range of audio frequencies. These devices are usually constructed using noninductive resistors, which helps to ensure a flat frequency response. Two examples of simple resistive *attenuator pads* are shown in Fig. 10-8. The circuit at A is a generic attenuator for balanced input and output. At B, a similar attenuator is shown for unbalanced input and output. The values of the resistors depend on the input impedance, the output impedance, and the extent to which the circuit attenuates the signal. The circuits shown are for *fixed attenuators*. That means they cause the same number of decibels of attenuation under all conditions. For a *variable attenuator* in which the number of decibels of attenuation can be adjusted, the design is rather complicated if constant input and output impedance are to be maintained.

All attenuators must be built to withstand the power applied to them. In attenuators designed for power amplifiers, the components must dissipate considerable power, sometimes hundreds of watts. Large noninductive resistors, or series-

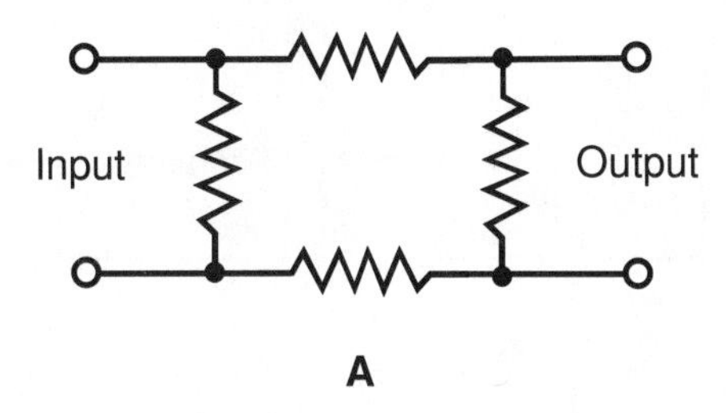

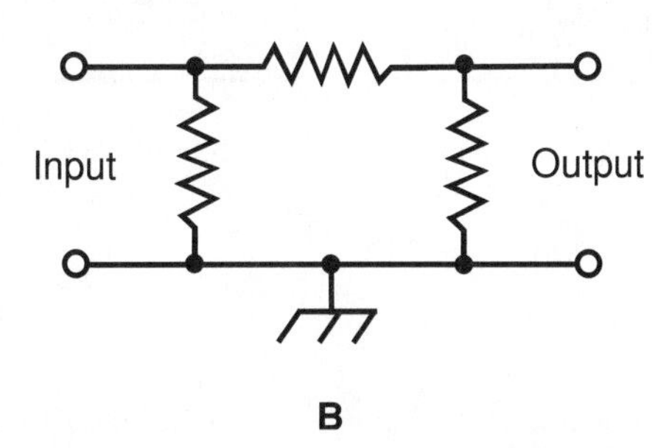

Figure 10-8 Basic attenuators composed of resistors. At A, balanced; at B, unbalanced.

parallel combinations of smaller resistors, are used to obtain the needed power-dissipation rating.

AUTOMATIC GAIN CONTROL

In some audio amplifier systems, it is advantageous to regulate the gain automatically, increasing it when the input signal amplitude drops, and decreasing it when the input signal amplitude rises. In this way, the difference between minimum and maximum output can be reduced as desired. This feature, called *automatic gain control* (AGC), allows low-level sounds to be heard while *blasting* (excessive volume) on signal peaks is prevented. It is also known as *automatic volume control* (AVC) or *automatic level control* (ALC).

Figure 10-9 is a simplified block diagram of an AGC system. There are two stages of amplification. Part of the output from the second stage is rectified, producing a pulsating DC voltage. This is filtered so the audio waveform component is eliminated, leaving a pure DC voltage that varies smoothly along with the volume of the audio input signal. The stronger the signal, the higher the voltage. A delay circuit, which can be as simple as a switch-selectable set of resistors and

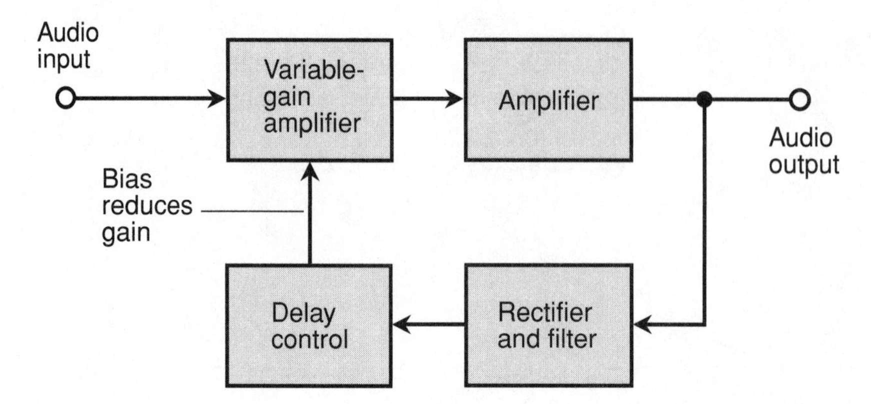

Figure 10-9 Block diagram of an automatic gain control (AGC) system in audio amplification.

capacitors of various values, makes it possible for the user to adjust the speed with which the DC voltage decays after an audio signal peak. This circuit has a *time constant* that produces some lag in the DC voltage with respect to the signal. The extent of the lag is called the *release time*. In contrast to this, the *attack time* is the time required for the DC voltage to rise to its maximum when a strong signal appears at the amplifier output. It is nearly always desirable for the attack time to be short.

The output of the delay circuit, a varying DC voltage, is injected into the first amplifier so as to reduce its gain. As the strength of the signal at the output of the second amplifier increases, the gain of the first amplifier decreases. Therefore, large variations in input signal amplitude result in proportionately smaller variations in the output signal amplitude. In the extreme, AGC makes it possible to maintain a constant (although amplified) output amplitude for a wide range of input amplitudes.

In audio recording equipment, an AGC circuit can be connected in the microphone amplifier chain. The AGC makes it possible, for example, to record the voice of a student speaking in a classroom even if he or she is on the other side of the room provided the ambient noise level in the room is low. The teacher can interject comments at close range without having to worry about blasting when the recording is played back. Automatic gain control can also be used in hi-fi sound systems. It is usually in the same set of audio amplifiers as the manual volume controls. The AGC can be switched off if it is not wanted. With the AGC activated, extreme sound peaks can be prevented from causing distortion.

PROBLEM 10-7

What is *peak clipping*? How does it differ from flat topping? What adverse effects can result from peak clipping?

SOLUTION 10-7

Peak clipping is an exaggerated form of flat topping. The wave peaks are sharply cut off, producing points in the waveform where the slope changes abruptly. An example of this is shown in Fig. 10-10. At A, the output signal waveshape for a properly operating class-A or class-B push-pull audio amplifier is shown, when a pure sine wave is applied to the input. At B, the output of the same amplifier is shown under conditions of extreme overdrive, again with a pure sine wave input. In hi-fi applications, the adverse effects of peak clipping are similar to those of flat topping, but worse. In particular, harmonic generation is more pronounced, to the extent that the output sound distortion is obvious even to untrained listeners.

PROBLEM 10-8

What happens in an audio system where the amplifier chain does not exhibit a flat gain-vs.-frequency function? How can problems that result from this be overcome?

SOLUTION 10-8

When the gain-vs.-frequency function of an amplifier is not flat, the resulting sound depends on various factors. The speakers are particularly important here. High-quality speakers, capable of reproducing sound over the widest possible range, are an absolute must in any high-end system. If the overall system gain is

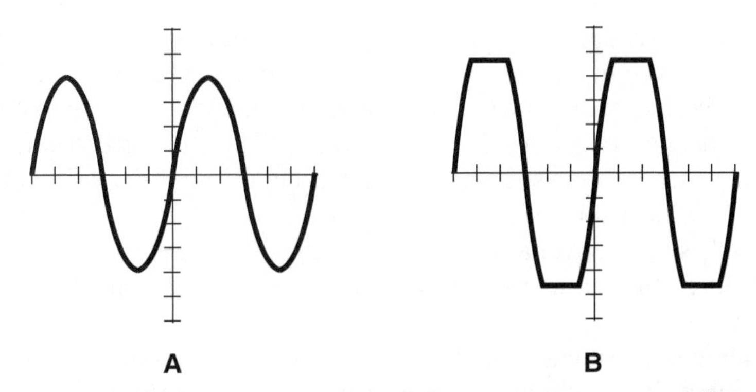

Figure 10-10 At A, the output of an amplifier with an input sine wave of the proper amplitude. At B, the output of the same amplifier with clipped peaks, resulting in extreme distortion.

greater at the bass frequencies than at higher frequencies, and if the speakers are ideal (flat response), the output will sound "boomy" or "muffled." If the gain is greater at the treble frequencies than at lower frequencies, the output will sound "crisp" or "tinny." To some extent, these phenomena occur in all systems, and they can be mitigated by the judicious use of tone controls. However, the best way to deal with problems of this sort, and to tailor the sound to suit the listener's preference as well as the acoustics of the room in which a system is installed, is to employ *audio filters* and/or *equalizers* in both channels. This subject is dealt with in Chapter 11.

Oscillator Fundamentals

An *oscillator* is an amplifier with significant *positive feedback*. Some of the output of an amplifier is applied to its own input, in phase with the original input. The slightest stimulus will set the circuit to going in a "vicious circle" where energy goes from the input through the amplifier to the output, back to the input and out again, endlessly. Uncontrolled *oscillation* can take the form of rumbling, howling, or screeching. By controlling the frequencies at which the feedback occurs, the output waveform of an oscillator can be tailored to any shape desired, producing musical notes that are pleasing to the ear. That is the basis for the operation of a *music synthesizer*.

POSITIVE FEEDBACK

For a circuit to oscillate, the fed-back output signal must be in phase, or nearly in phase, with the original input signal. This is what is meant by the term positive feedback. The fed-back signal should be strong enough to cause and maintain oscillation, but not too strong. The ideal amount of positive feedback is the minimum amount necessary to sustain oscillation under all the operating conditions the circuit is likely to encounter.

Out-of-phase (negative) feedback will not cause oscillation. It reduces the gain of an amplifier. In fact, negative feedback is used in some amplifiers to *prevent* oscillation.

The output of a common-emitter or common-source amplifier is out of phase from the input. If you couple the collector or drain directly to the base or gate through a large-value series capacitor, you won't get oscillation. It is necessary to reverse the phase in the feedback process in order for oscillation to occur in a

common-emitter or common-source circuit. In addition, the amplifier gain must be high, and the coupling from the output to the input must be good. The positive feedback path must be easy for a signal to follow.

The output of a common-base or common-gate amplifier is in phase with the input. However, these circuits have limited gain, and it's hard to make them oscillate. Common-collector and common-drain circuits don't have enough gain to work well as oscillators.

THE ARMSTRONG OSCILLATOR

A common-emitter or common-source class-A amplifier can be made to oscillate by coupling the output back to the input through a transformer that reverses the phase of the fed-back signal. Figure 10-11 shows a common-source N-channel JFET amplifier with the drain circuit coupled to the gate circuit by means of a transformer.

The oscillation frequency is determined by a capacitor in series with the secondary winding of the transformer. The inherent inductance L of the transformer secondary, along with the capacitance C, forms a *series resonant circuit* that passes energy easily at one frequency while attenuating the energy at other frequencies.

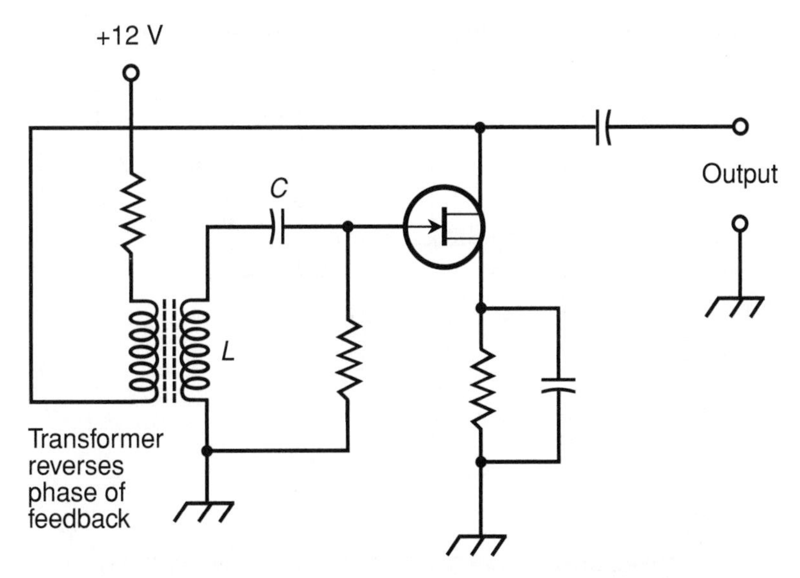

Figure 10-11 Armstrong oscillator circuit using an N-channel JFET. The oscillation frequency is determined by the inductance L and the capacitance C.

This circuit is known as an *Armstrong oscillator*. A bipolar transistor can be used in place of the JFET, as long as the device is biased for class-A amplification.

THE TWIN-T OSCILLATOR

A popular circuit for general-purpose audio signal generation is the *twin-T oscillator* (Fig. 10-12). The frequency is determined by the resistances *R* and capacitances *C*. The output is a near-perfect sine wave, but it's not quite perfect. The small amount of distortion helps to alleviate the irritation produced by an absolutely pure sinusoid. The circuit shown in this example uses two PNP bipolar transistors biased for class-A amplification.

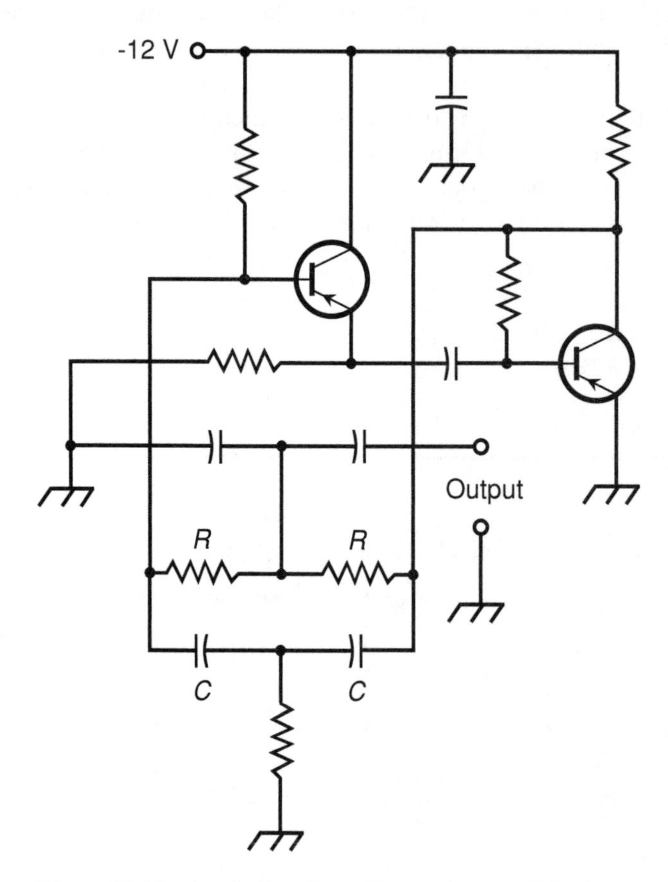

Figure 10-12 A twin-T audio oscillator using two PNP bipolar transistors. The frequency is determined by the resistances *R* and the capacitances *C*.

THE "MULTIVIBRATOR"

Another type of audio oscillator circuit makes use of two identical common-emitter or common-source amplifier circuits, hooked up so that the signal goes around and around between them. This is sometimes called a *"multivibrator"* (that is technically a misnomer, the term being more appropriate to various digital signal-generating circuits).

In the example of Fig. 10-13, two N-channel JFETs are connected to form an AF "multivibrator" circuit. Each JFET amplifies the signal in class-A, and reverses the phase by 180°. Therefore, the signal goes through a 360° phase shift each time it gets back to any particular point. A 360° phase shift is equivalent to no phase shift, so it results in positive feedback.

The frequency of the circuit shown in Fig. 10-13 is set by means of an inductance-capacitance (*LC*) circuit. The coil uses a ferromagnetic core, because stability is not of great concern and because such a core is necessary to obtain the large inductance needed for resonance at AF. *Toroidal cores* or *pot cores* are excellent in this application.

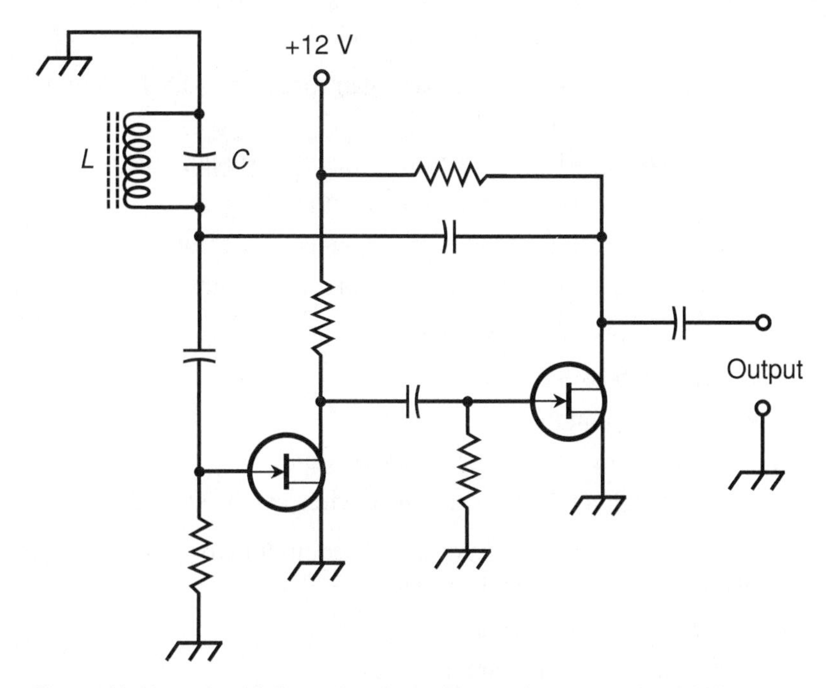

Figure 10-13 A "multivibrator" audio oscillator using two N-channel JFETs. The frequency is determined by the inductance *L* and the capacitance *C*.

PROBLEM 10-9

In the circuit of Fig. 10-12, what would happen if the connection between the emitter of the right-hand transistor and ground were to open up?

SOLUTION 10-9

This would cause failure of the second amplifier, and the oscillator would stop working.

PROBLEM 10-10

In the circuit of Fig. 10-13, what would happen if the capacitance *C* on the left-hand side were replaced by a direct short circuit?

SOLUTION 10-10

This would ground the gate of the left-hand transistor for the signal (although not for DC). The result would be to deprive the left-hand transistor of its input, interrupting the feedback loop and causing failure of the oscillator.

Quiz

This is an "open book" quiz. You may refer to the text in this chapter. A good score is 8 correct. Answers are in the back of the book.

1. Suppose a circuit is found to have a gain figure of –15 dB. Which of the following statements is true?
   (a) The output signal is stronger than the input signal.
   (b) The input signal is stronger than the output signal.
   (c) The input signal is 15 times as strong as the output signal.
   (d) The output signal is 15 times as strong as the input signal.
2. In an oscillator circuit, the feedback should be
   (a) as great as possible.
   (b) kept to a minimum.
   (c) just enough to reliably sustain oscillation.
   (d) done through a transformer whose wires can be switched easily.
3. Which type of bipolar-transistor amplifier circuit works best as an oscillator when positive feedback is provided?
   (a) The common-base circuit.
   (b) The common-emitter circuit.
   (c) The common-collector circuit.
   (d) Any of the above amplifier arrangements will work fine as the basis for

an oscillator using a bipolar transistor.

4. In which type of oscillator does a transformer provide the feedback?
   (a) The common-base oscillator.
   (b) The twin-T oscillator.
   (c) The multivibrator.
   (d) The Armstrong oscillator.
5. Fill in the blank to make the following sentence true: "The efficiency, in percent, of a JFET audio power amplifier is equal to 100 times the ratio of the ________ to the DC drain power input."
   (a) DC source power input
   (b) audio power output
   (c) audio power input
   (d) DC gate power output
6. The arrangement in the block diagram of Fig. 10-14 represents
   (a) a method of testing an amplifier for distortion.
   (b) a method of controlling the frequency of an audio oscillator.
   (c) a device for generating an audio signal.
   (d) a circuit that can eliminate flat topping and peak clipping.
7. In which of the following JFET amplifier types does drain current flow even in the absence of an input signal?
   (a) The class-B circuit.
   (b) The common-source circuit.
   (c) The common-drain circuit.

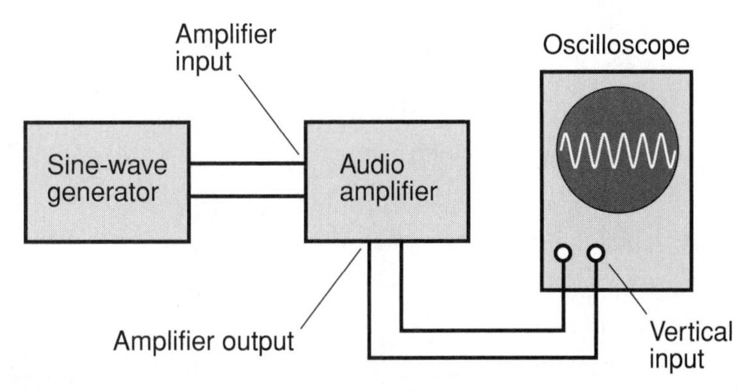

Figure 10-14 Illustration for Quiz Question 6.

(d) The class-A circuit.

8. Which bipolar amplifier type has a distortion-free output signal waveform in the collector circuit of a single transistor?
   (a) The class-B circuit.
   (b) The common-emitter circuit.
   (c) The common-collector circuit.
   (d) The class-A circuit.

9. Suppose a bipolar-transistor amplifier is 66 percent efficient. The output power is 33 W. The DC collector power input is
   (a) 22 W.
   (b) 50 W.
   (c) 2.2 W.
   (d) impossible to determine without more information.

10. Class-B amplification be used to obtain low distortion for audio applications
    (a) by connecting two amplifiers in cascade (one after the other), thereby maximizing the gain.
    (b) by biasing the bipolar transistor or FET beyond cutoff or pinchoff, thereby ensuring that the output in phase with the input.
    (c) by connecting two identical bipolar transistors or FETs, biased exactly at cutoff or pinchoff, in a push-pull configuration.
    (d) by biasing the bipolar transistor or FET in the middle of the straight-line portion of the characteristic curve.

Filters and Equalizers

In this chapter, you'll learn how electronic devices and circuits can be used to tailor the frequency response of an audio system. This is how recording engineers—and you—can get the audio to sound exactly as you want.

Meet the Op Amp

An *operational amplifier*, more often called an *op amp*, is a specialized *integrated circuit* (IC) that consists of several bipolar transistors, resistors, diodes, and capacitors, interconnected so that amplification can be achieved over a wide range of frequencies. Some ICs consist of two or four individual op amps, so you'll sometimes hear or read about *dual op amps* and *quad op amps*. Some ICs have op amps in addition to other circuits, fabricated onto a single *chip* of semiconductor material.

TERMINALS

An op amp has two input terminals, one *non-inverting* (denoted by a plus sign) and one *inverting* (denoted by a minus sign). There is one output terminal. When a sig-

nal goes into the non-inverting input, the output wave is in phase with the input wave. When a signal goes into the inverting input, the output wave is "upside-down," or inverted, with respect to the input wave. There are two power supply connections, one for the emitters of the transistors (V_{ee}) and one for the collectors (V_{cc}). The usual schematic symbol for an op amp is a triangle. The inputs, output, and power supply connections are drawn as lines emerging from this triangle, as shown in Fig. 11-1.

THEORETICAL CHARACTERISTICS

A perfect op amp would have infinite, purely resistive input impedance and infinite bandwidth, so it would appear as an open circuit to input signals at all possible frequencies. It would also have zero output impedance and infinite gain. These theoretical ideals are, of course, never achieved in practice. Nevertheless, a well-designed op amp has very high input impedance, so it draws practically no power from the signal source. The output impedance is extremely low, and the maximum achievable gain is extremely high—more than 100 dB in some cases—and the bandwidth can range over several orders of magnitude, from well below the AF range up to several tens of megahertz. Op amps are especially well suited for use in devices that modify the characteristics of AF energy.

OPEN-LOOP CONFIGURATION

An op amp has the highest possible gain when running "wide open," that is, in the *open-loop configuration*. Figure 11-2A is a diagram of a simple, broadband open-loop op-amp circuit in which the output is in phase with the input. This is called a

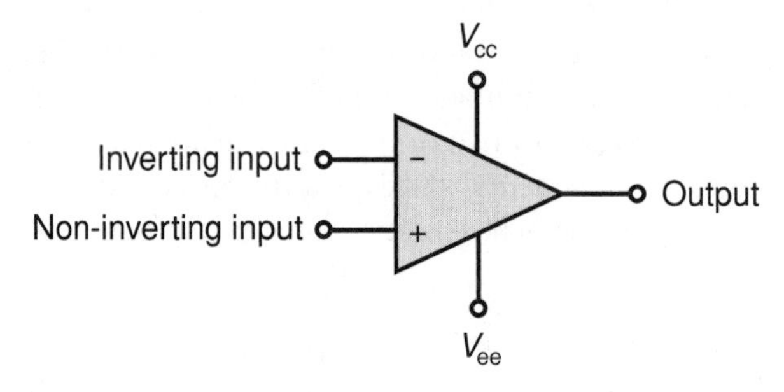

Figure 11-1 Schematic symbol and connections for an operational amplifier (op amp).

non-inverting amplifier because it does not reverse the instantaneous polarity of the input signal, which is applied to the non-inverting (+) input terminal. Figure 11-2B is a diagram of a basic broadband open-loop op-amp circuit in which the instantaneous output polarity is reversed with respect to the instantaneous input polarity. That is, the output signal is "upside down" from the input. This is an example of an *inverting amplifier*. The input signal is applied to the inverting (–) input terminal.

CLOSED-LOOP CONFIGURATION

An op amp nearly always has more gain than is necessary or desirable when used in the open-loop configuration. To control the gain, *negative feedback* is introduced. This feedback is obtained by connecting the output to the inverting input through a resistor, potentiometer, or a network of various components.

An op-amp circuit with negative feedback is said to be functioning in the *closed-loop configuration*. This stabilizes the input impedance of the circuit at a finite value, and also evens out the circuit performance over a wide band of frequencies.

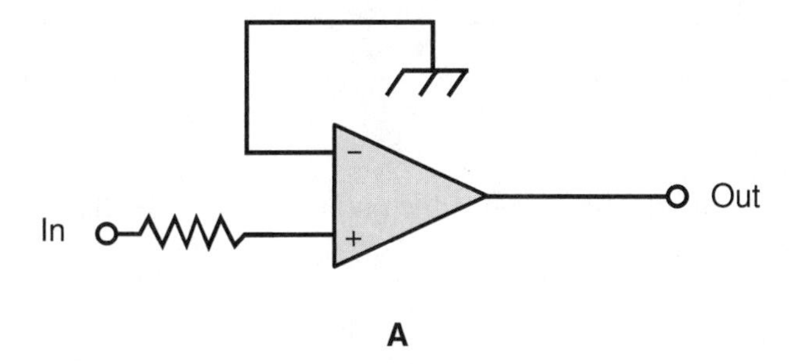

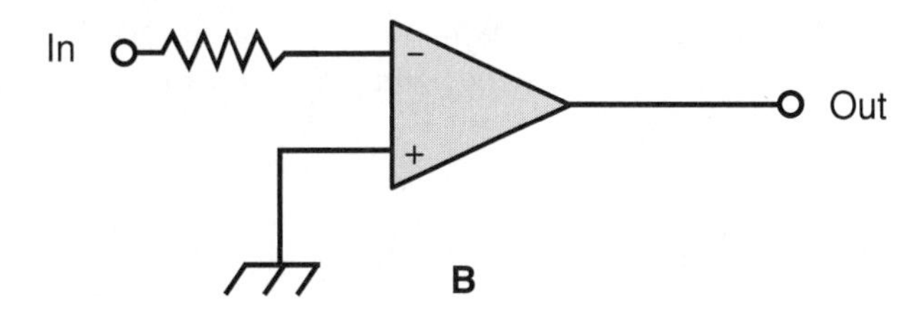

Figure 11-2 At A, a non-inverting, open-loop op-amp circuit. At B, an inverting, open-loop op-amp circuit. Power-supply connections are not shown; this omission is common in schematic diagrams involving op amps.

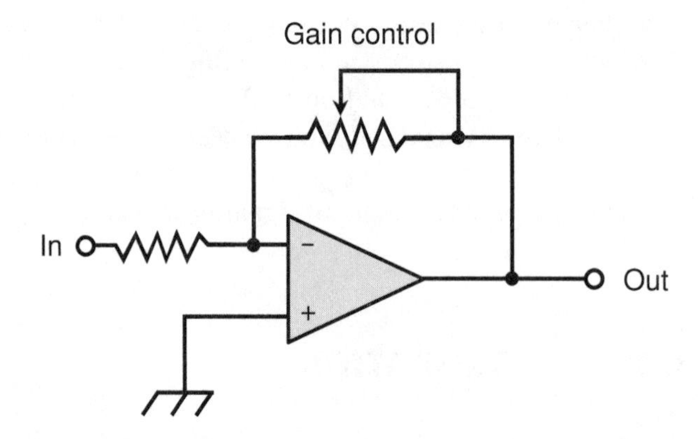

Figure 11-3 A closed-loop op-amp circuit incorporating adjustable negative feedback using a potentiometer to vary the gain. The gain is directly proportional to the resistance of the potentiometer.

Negative feedback can allow adjustable gain, such as in the volume control for a high-fidelity amplifier. Figure 11-3 is a schematic diagram of an op-amp circuit with adjustable negative feedback to vary the gain. The gain is directly proportional to the feedback-loop resistance. As the potentiometer resistance goes down, so does the gain.

PROBLEM 11-1

What happens in an op-amp circuit if the output is connected through a resistor to the non-inverting input?

SOLUTION 11-1

This produces positive feedback, and can cause oscillation. If a resistance-capacitance (*RC*) combination is used, the frequency of the oscillation can be programmed. By carefully choosing the resistance(s) and capacitance(s) and by "tweaking" amount of feedback, the shape of the generated waveform can be tailored. The result is a circuit called a *waveform generator* or *function generator*.

The Active Audio Filter

When *RC* combinations are used along with one or more op amps, the result is an *active audio filter*. When an *RC* combination is used in the feedback loop of an op amp from the output to the inverting input, an amplifier results, with the gain

dependent on the input-signal frequency. Circuits of this type form the basis for sophisticated audio tone controls and equalizers.

INSERTION POWER GAIN

Insertion power gain is a comparison of output signal power, in decibels (dB), with and without an op-amp filter in the line when the input frequency is within the range where the filter has the most gain (or the least attenuation). If P_1 is the signal output power without the filter in the line and P_2 is the signal output power with the filter in the line, then the *insertion power gain*, G_P, is:

$$G_P = 10 \log (P_2/P_1)$$

ULTIMATE POWER ATTENUATION

Ultimate power attenuation is a quantitative expression of the maximum extent to which a filter can block signals at unwanted frequencies far removed from desired frequencies, relative to the output amplitude at the desired frequencies. Let U_P represent the ultimate power attenuation in decibels, let P_d represent the output signal power at desired frequencies, and let P_n represent the output signal power at frequencies that are not desired. Then:

$$U_P = 10 \log (P_d/P_n)$$

Both the insertion power gain and the ultimate power attenuation figures are specified in decibels. It is assumed that both power figures are specified in the same units (for example, watts or milliwatts). It is also assumed that the input power remains constant.

Note that U_P is defined with respect to the maximum unattenuated *output* power, and not with respect to the input power.

GAIN-VERSUS-FREQUENCY CHARACTERISTIC

The *gain-vs-frequency characteristic*, also called the *gain-vs.-frequency curve*, for a filter circuit is a graph representing the amount of power amplification in decibels (positive or negative) through the circuit as a function of frequency. This graph, which appears as a mathematical function, is shown with the amplitude (in decibels relative to a certain reference level) on the vertical scale, and the frequency on the horizontal scale.

PROBLEM 11-2

Suppose a certain audio filter is provided with 10 mW of audio input power. At desired frequencies, the output power is 1.5 W. At unwanted frequencies far removed from the desired frequencies, the output power is 1.5 mW. What is the ultimate power attenuation?

SOLUTION 11-2

Convert both output power figures to the same units. Let's use milliwatts. Then we have $P_\mathrm{d} = 1500$ mW and $P_\mathrm{n} = 1.5$ mW. Using the above formula:

$$\begin{aligned} U_P &= 10 \log (P_\mathrm{d}/P_\mathrm{n}) \\ &= 10 \log (1500/1.5) \\ &= 10 \log 1000 \\ &= 10 \times 3 \\ &= 30 \text{ dB} \end{aligned}$$

Basic Filter Configurations

It is possible to get a *lowpass response*, a *highpass response*, a *bandpass response*, or a *band-rejection response* using an op amp and various *RC* feedback arrangements. These four types of gain-vs.-frequency characteristics are shown qualitatively in Fig. 11-4.

ACTIVE LOWPASS FILTER

An *active lowpass filter* is intended to produce power gain below a certain critical frequency, and less gain (or even a loss) above that frequency. At the critical frequency, the power gain of the circuit is –3 dB with respect to the *maximum gain* (also known as the *minimum attenuation*). Another way of saying this is that the power attenuation (or *power loss*), relative to the maximum circuit gain, is 3 dB. For frequencies at or below the critical frequency, the attenuation is less than or equal to 3 dB. For frequencies above the critical frequency, the attenuation is more than 3 dB.

The 3-dB power attenuation figure is used by convention, because a reduction of 3 dB represents a 50 percent decrease in signal power. In a filter circuit, any frequency at which the output signal power is 3 dB down relative to the output signal power at the frequency or frequencies of maximum gain is known as a *half-*

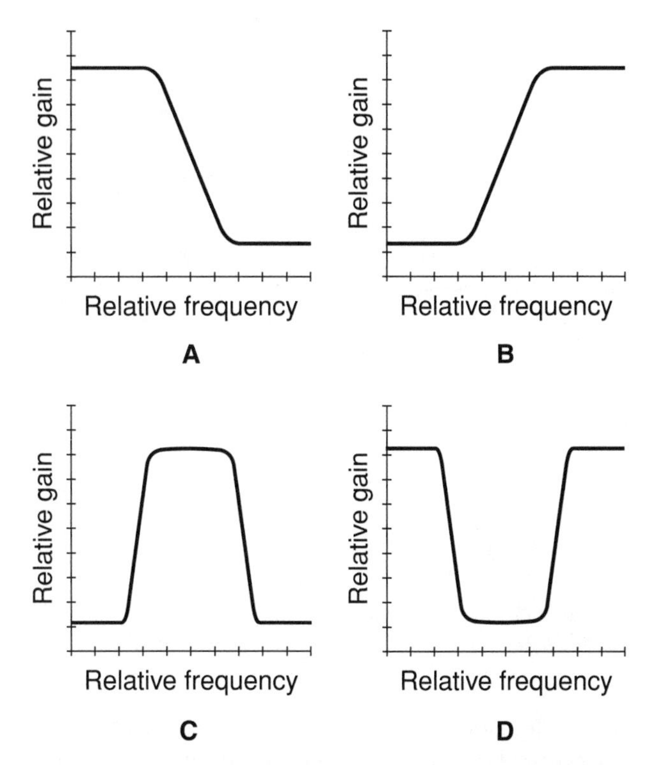

Figure 11-4 Gain-vs.-frequency response curves. At A, lowpass; at B, highpass; at C, bandpass; at D, band-rejection.

power point. The span, or range, of frequencies at which the attenuation is less than or equal to 3 dB is called the *passband*. The span, or range, of frequencies at which the attenuation is greater than 3 dB is called the *stopband*.

Figure 11-5 is a schematic diagram of a generic active lowpass filter using an op amp. The half-power point is determined by the resistance R and the capacitance C. If a potentiometer is used in place of the resistor, the half-power point can be adjusted over a continuous range.

ACTIVE HIGHPASS FILTER

An *active highpass filter* produces power gain above a certain frequency, and less gain (or even a loss) below that frequency. At the half-power point, the power gain of the circuit is –3 dB with respect to the maximum gain. For frequencies at or above the half-power point, the attenuation is less than or equal to 3 dB. For frequencies below the half-power point, the attenuation is more than 3 dB.

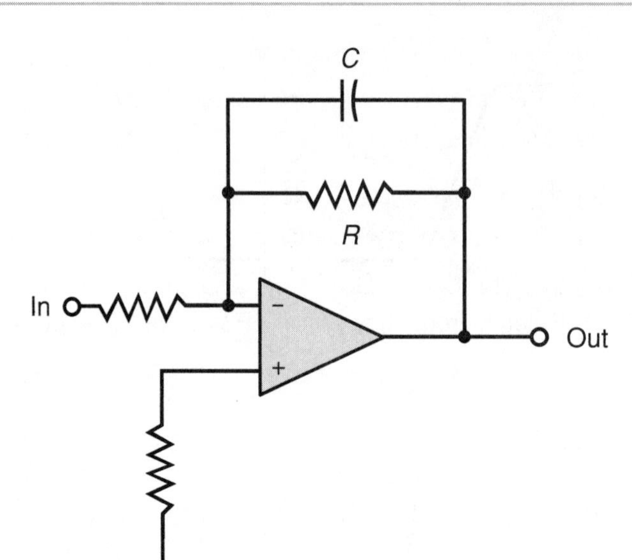

Figure 11-5 An active audio lowpass filter. The half-power point is determined by the resistance *R* and the capacitance *C*.

Figure 11-6 is a schematic diagram of a generic active highpass filter using an op amp. The half-power point is determined by the resistance *R* and the capacitance *C*. If a potentiometer is used in place of the resistor, the half-power point can be adjusted over a continuous range.

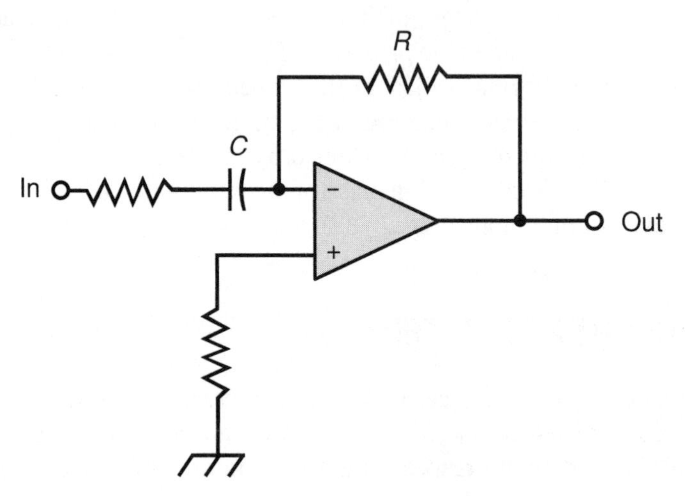

Figure 11-6 An active audio highpass filter. The half-power point is determined by the resistance *R* and the capacitance *C*.

ACTIVE BANDPASS FILTER

An *active bandpass filter* produces power gain in a defined range, or *band*, between two specific frequencies, and less gain (or even a loss) at frequencies outside that band. There are two half-power points at which the power gain of the circuit is –3 dB with respect to the maximum gain. For frequencies at or between these points, the attenuation is less than or equal to 3 dB. At frequencies above the upper half-power point or below the lower half-power point, the attenuation is greater than 3 dB.

Figure 11-7 is a simplified diagram of a generic active bandpass filter. In this design, a lowpass filter and a highpass filter are *cascaded* (connected in series) with an isolation circuit between them to prevent unwanted impedance interaction between the op amps. The passbands of the two filters must overlap.

In the band of frequencies where the passbands overlap, audio signals can get through both filters with ease. But outside that band, one or the other of the filters impedes the signal. If potentiometers, rather than fixed resistors, are included in the lowpass and highpass filters, the lower and upper half-power points can be independently adjusted. This varies the span of frequencies over which the filter will easily pass AF signals.

ACTIVE BAND-REJECTION FILTER

An *active band-rejection filter*, also known as an *active band-stop filter*, produces power gain outside of a defined frequency band, and less gain (or even a loss) within that band. As with the bandpass filter, there are two half-power points at which the power gain of the circuit is –3 dB with respect to the maximum gain. But the gain-vs.-frequency characteristic is "upside down" compared to that of a bandpass filter. For frequencies at or between the half-power points, the attenuation

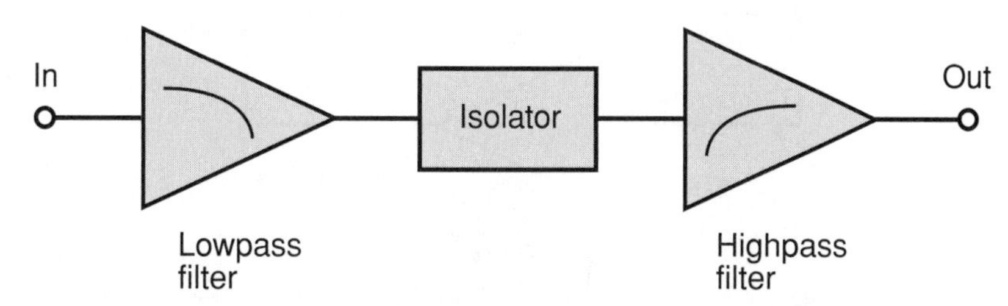

Figure 11-7 An active audio bandpass filter can be designed by cascading a lowpass filter and a highpass filter, with isolation in between them to prevent interaction.

is greater than or equal to 3 dB. At frequencies above the upper half-power point or below the lower half-power point, the attenuation is less than 3 dB.

Figure 11-8 is a simplified diagram of a generic active band-rejection filter. In this design, a lowpass filter and a highpass filter are connected in parallel. A *splitter* is used at the input, and a *combiner* is used at the output. These circuits prevent unwanted impedance interaction between the op amps. The passbands of the two filters are disjoint. That means the stopbands, rather than the passbands, overlap. The half-power point of the lowpass filter is below the half-power point of the highpass filter.

In the band of frequencies where the stopbands overlap, audio signals are significantly attenuated by both of the filters. Outside that band, one or the other of the filters lets the signal through with comparative ease. Again, as with the passband filter, potentiometers can be used rather than fixed resistors in the lowpass and highpass filters. That way, the lower and upper half-power points can be independently adjusted. This varies the span of frequencies over which the filter will significantly attenuate AF signals.

PROBLEM 11-3

Suppose, in the active bandpass filter design described above, the half-power points for the lowpass and highpass filters are identical. What will the gain-vs.-frequency characteristic of the resulting filter look like?

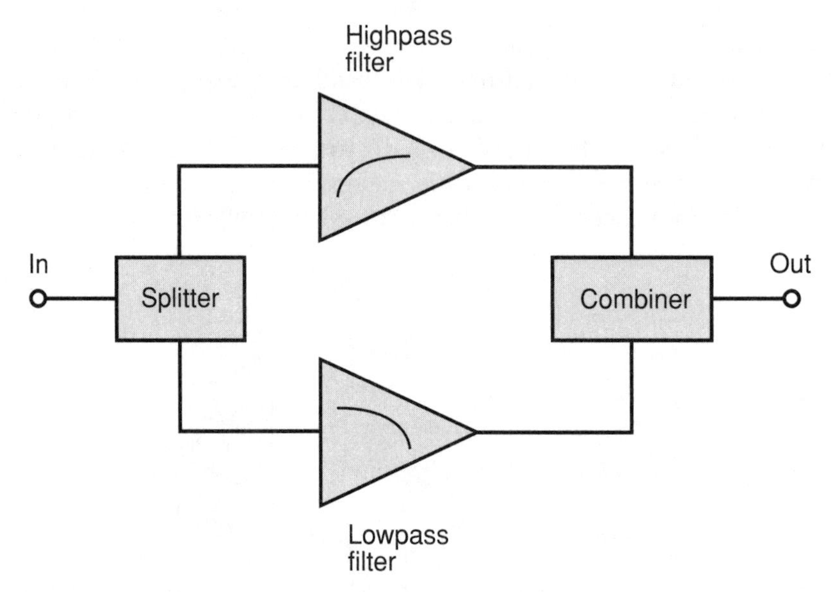

Figure 11-8 An active audio band-rejection filter can be designed by connecting a lowpass filter and a highpass filter in parallel, with a splitter and combiner to prevent interaction.

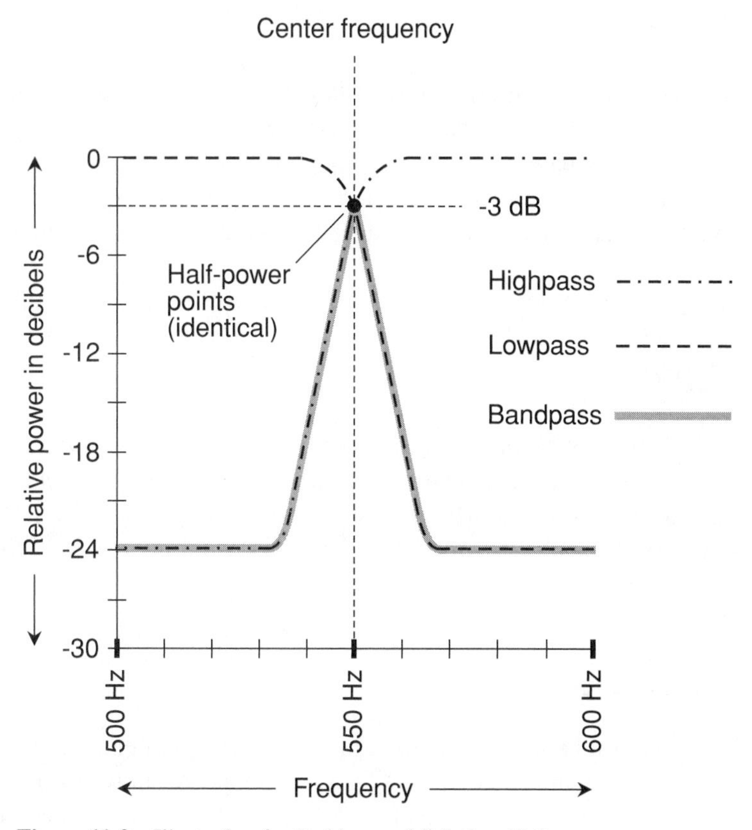

Figure 11-9 Illustration for Problem and Solution 11-3.

SOLUTION 11-3

The filter will exhibit a sharp peak at a single frequency, with less gain above and below that frequency, as illustrated in Fig. 11-9. The highpass and lowpass responses are shown as dashed curves; the bandpass response is shown as a heavy gray curve. A bandpass characteristic of this sort is known as a *peaked response*.

Two Important Parameters

Two more variables, also called *parameters*, can be specified for audio filters. They are the *bandwidth* and the *skirt slope*.

BANDWIDTH

In an audio bandpass or band-rejection filter, the bandwidth is a quantitative expression of the "sharpness" of the gain-vs.-frequency characteristic. Suppose the half-power points in a bandpass or band-rejection filter are represented by frequencies, in hertz, of f_0 (the lower of the two) and f_1 (the higher of the two). The filter bandwidth in hertz, B_{Hz}, is defined as the difference between them:

$$B_{Hz} = f_1 - f_0$$

SKIRT SLOPE

Skirt slope is an expression of the abruptness of the transition between the passband and the stopband of a filter. The steeper the slope, the more abrupt the transition. Lowpass and highpass filters have only one such transition (or *skirt*); bandpass and band-rejection filters have two. Engineers can design audio filters that have adjustable skirt slopes.

A HYPOTHETICAL FILTER

Figure 11-10 is a graphical illustration of the gain-vs.-frequency characteristic for a hypothetical bandpass filter. The half-power points are shown as dots plotted at the frequencies where the filter gain is –3 dB with respect to the minimum attenuation.

The frequency at the middle of the passband, which is the average of the half-power points f_0 and f_1, is called the *center frequency*. In this example, suppose $f_0 = 531$ Hz and $f_1 = 569$ Hz. Then the center frequency, f_c, is:

$$\begin{aligned} f_c &= (f_0 + f_1)/2 \\ &= (531 + 569)/2 \\ &= 550 \text{ Hz} \end{aligned}$$

The bandwidth, in hertz, is the difference between the upper and lower half-power points in hertz:

$$\begin{aligned} B_{Hz} &= f_1 - f_0 \\ &= 569 - 531 \\ &= 38 \text{ Hz} \end{aligned}$$

The ultimate power attenuation for this filter can be read directly from the vertical scale, which denotes the relative power in decibels. For frequencies well

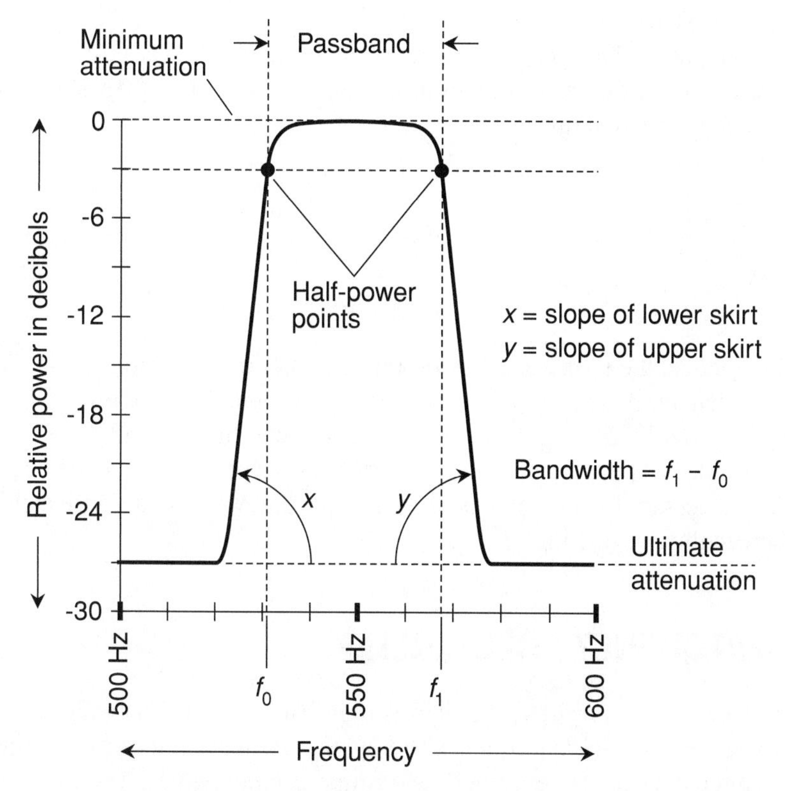

Figure 11-10 Graphical illustration of parameters for a hypothetical bandpass filter centered near 550 Hz.

removed from the passband, the gain of this filter is –27 dB with respect to the maximum gain.

Skirt slope is often defined in a subjective way, using expressions such as "steep" or "shallow." But if the skirts appear as straight lines (as they do in this example), the slopes for the lower and upper skirts can be specified as angles in degrees. These angles, shown as x and y in the illustration, are always larger than 0° and less than 90°. The smaller the angle, the "shallower" the skirt. If both the lower and upper skirt slopes are close to 90° in a bandpass or band-rejection filter, and if the passband exhibits a flat (rather than a peaked) response over a significant range of frequencies, then that filter is said to have a *rectangular response*.

PROBLEM 11-4

Is it possible for the lower and upper skirt slopes in a bandpass or band-rejection filter to differ?

SOLUTION 11-4

Yes. If a bandpass or band-rejection filter is made up of a lowpass filter and a highpass filter that have different skirt slopes, then the lower and upper skirt slopes in the composite filter will differ.

Equalization

In audio practice, the term *equalization* refers to any modification of the gain-vs.-frequency characteristic that is done with the intent of tailoring the way the output sounds. This can be done for esthetic reasons alone, but it is often necessary in order to compensate for the acoustical characteristics of the environment. In some cases, it is necessary to prevent *acoustic feedback* that would otherwise cause rumbling, screeching, or howling.

LOW-FREQUENCY SHELF FILTER

In a *low-frequency shelf filter*, the gain is unity (that is, a power ratio of 1:1 or 0 dB) above a certain critical frequency. Below that frequency, the gain is adjustable and can range over negative as well as positive decibel values. Therefore, the filter can act either as a low-frequency amplifier or as a low-frequency attenuator, in addition to offering a flat response if desired.

Figure 11-11 shows a family of gain-vs.-frequency curves for a hypothetical low-frequency shelf filter. The *shelf frequency* is usually defined as the frequency at the center of the curve transition (the mid-point of the sloped portion). Rarely, it is defined as the frequency at which the gain begins to increase or decrease from its level in the flat-response range. The shelf frequency is adjustable. In some filters, the skirt slope, or rate of transition, can also be adjusted. You can think of a low-frequency shelf filter as a sophisticated form of bass level control.

HIGH-FREQUENCY SHELF FILTER

In a *high-frequency shelf filter*, the gain is unity below a certain critical frequency. Above that frequency, the gain can be continuously varied. The shelf frequency is also adjustable. The skirt slope can be varied in some devices. This type of filter can act either as a high-frequency amplifier or as a high-frequency attenuator, in addition to producing a flat response.

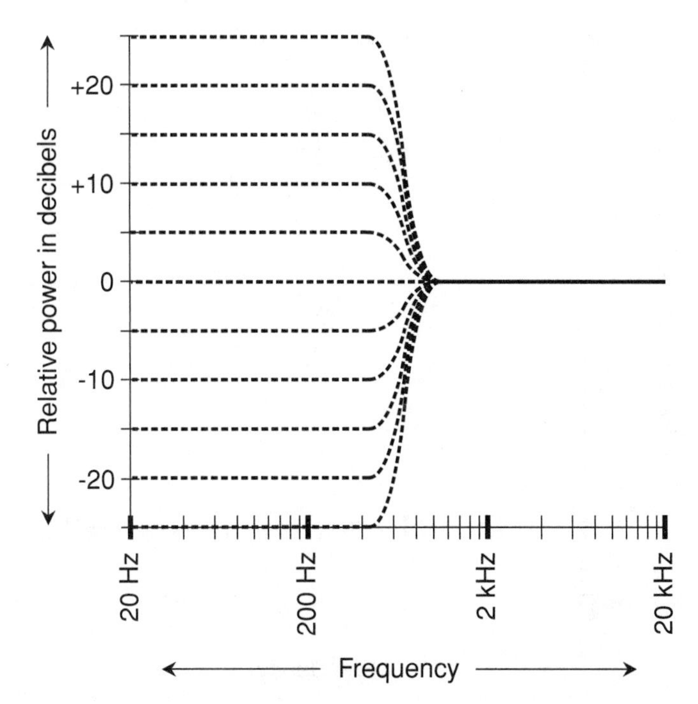

Figure 11-11 Some gain-vs.-frequency curves for a hypothetical low-frequency shelf filter. Dashed curves denote the shelving effect. The frequency scale is logarithmic.

Figure 11-12 shows a family of gain-vs.-frequency curves for a hypothetical high-frequency shelf filter. You might imagine it as a sophisticated treble level control. A low-frequency shelf filter and a high-frequency shelf filter, if combined, can act as a versatile tone control.

GRAPHIC EQUALIZER

A *graphic equalizer* is a device for adjusting the relative loudness of audio signals at various frequencies. It allows for meticulous tailoring of the amplitude-vs.-frequency output of hi-fi sound equipment. Equalizers are used in recording studios and by serious hi-fi stereo enthusiasts. There are several independent gain controls, each one affecting a different part of the audible spectrum. The controls are slide potentiometers with calibrated scales. The slides move up-and-down or left-to-right. When the potentiometers are set so that the slides are all at the same level, the audio output or response is flat, meaning that no particular range is amplified or attenuated with respect to the other ranges. By moving any one of the controls,

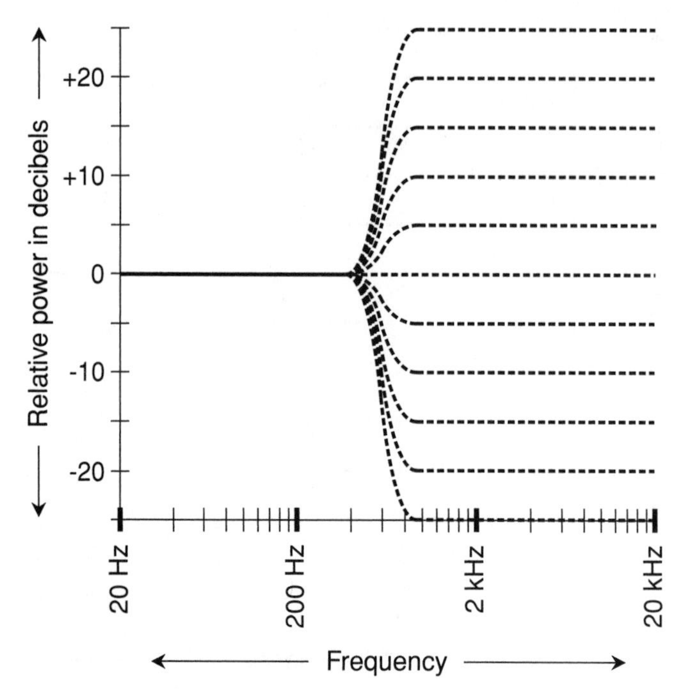

Figure 11-12 Some gain-vs.-frequency curves for a hypothetical high-frequency shelf filter. Dashed curves denote the shelving effect. The frequency scale is logarithmic.

the user can adjust the gain within a certain frequency range without affecting the gain outside that range. The positions of the controls on the front panel provide an intuitive graph of the output or response curve.

Figure 11-13 is a block diagram of a hypothetical graphic equalizer with six gain controls. The input is fed to a splitter that breaks the signal into six paths of equal impedance, and prevents interaction among the circuits. The six signals are fed to active audio bandpass filters, with each filter having its own gain control. The center frequencies can be spaced directly by frequency (that is, in a linear manner), or according to the logarithm of the frequency (as is the case in this example). The slide potentiometers are shown in Fig. 11-13 as variable resistors following the attenuators. Finally, the signals pass through an audio combiner, and the composite is sent to the output.

There are several challenges in the design and proper use of graphic equalizers. The individual gain controls must not interact. Judicious choice of filter frequencies and responses is important. The filters must not introduce distortion. The active devices must not generate significant audio noise. Graphic equalizers are not built to handle high power, so they must be placed at low-level points in an

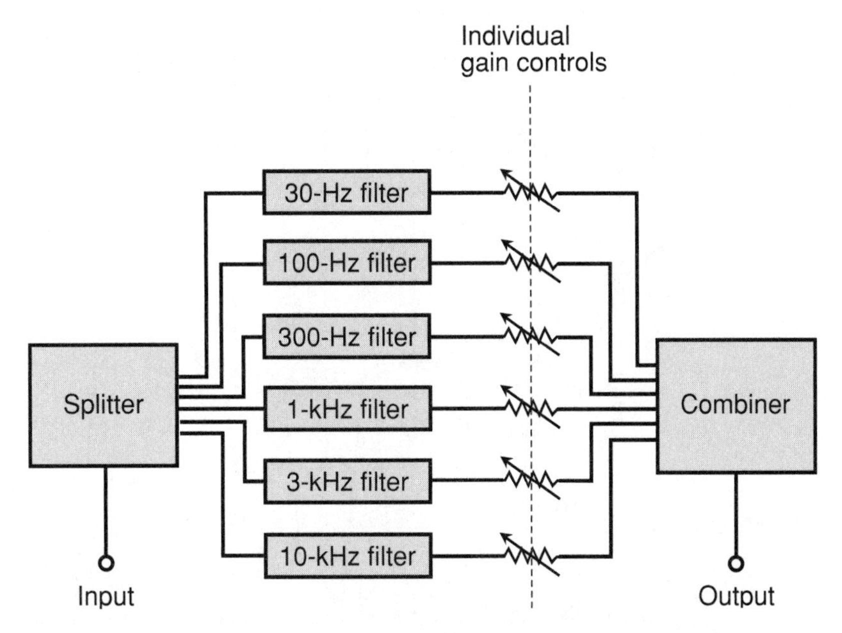

Figure 11-13 Block diagram of a graphic equalizer with six bandpass filters, equally spaced over the AF range according to the logarithm of the frequency.

audio amplifier chain. In a multichannel circuit such as a stereo sound system, a separate graphic equalizer can be used for each channel.

Figure 11-14 is an example of a gain-vs.-frequency characteristic (gray solid line) that can be obtained using the equalizer circuit diagrammed in Fig. 11-13. The individual bandpass responses are shown as dashed curves. This particular equalizer has six filters, but in high-end professional audio work, equalizers can have 20 or more individual filters for each audio channel in the system.

PARAMETRIC EQUALIZER

A graphic equalizer in which the gain, center frequency, bandwidth, and skirt slopes are independently adjustable for each filter is known as a *parametric equalizer*. In addition to several bandpass filters for each channel, a parametric equalizer can incorporate a low-frequency shelf filter and a high-frequency shelf filter, both of which have adjustable gain, shelf frequency, and skirt slope. A well-designed parametric equalizer can produce any gain-vs.-frequency characteristic an audiophile could possibly want, and can compensate for the acoustic characteristics of practically any reproduction or performance venue.

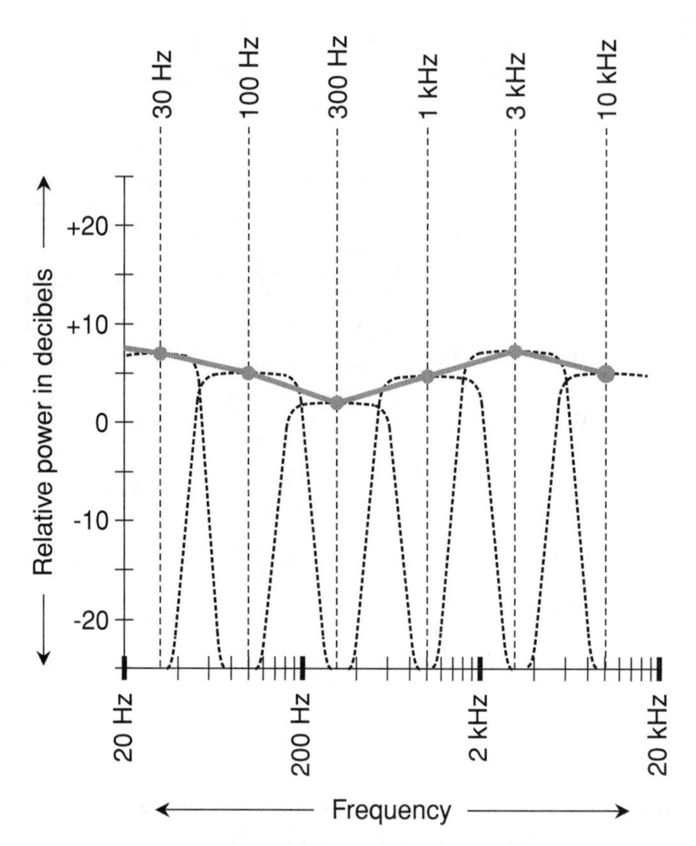

Figure 11-14 An example of a gain-vs.-frequency characteristic (gray solid line) that can be obtained using the bandpass filter diagrammed in Fig. 11-13. The individual bandpass responses are shown as dashed curves.

PROBLEM 11-5

Suppose an audio system is set up for a live performance in a certain auditorium. There is one problem: persistent feedback at a frequency of 853 Hz. Somehow, too much sound gets back from the speakers to the microphones at this frequency unless the system gain is turned down so low that the performance is adversely affected. The audio engineer suspects that the feedback has something to do with the dimensions or characteristics of the auditorium. He tries moving the speakers and microphones around, but none of these measures has the desired effect, and there is a limit to how much he can rearrange things without upsetting the band members. Fortunately, the band's equipment includes a good parametric equalizer. How can it be used to remedy the feedback problem?

SOLUTION 11-5

The equalizer can be adjusted so that one of the filters produces a high degree of attenuation at, and very close to, 853 Hz. This is done by creating a sharp band-rejection response, also known as a *notch*, at this frequency. A band-rejection filter designed or adjusted to attenuate AF at one specific frequency, while passing or amplifying signals at other frequencies, is called a *notch filter*. To obtain the notch in this scenario, one of the filters should be centered at 853 Hz, the gain set to minimum (or the attenuation to maximum), the lower and upper skirt slopes made steep, and the bandwidth set to the smallest value that prevents the feedback. The other filters are set for optimum sound quality by trial and error. The notch at 853 Hz will cause little or no noticeable degradation in the overall fidelity, but it should eliminate the feedback problem.

Quiz

This is an "open book" quiz. You may refer to the text in this chapter. A good score is 8 correct. Answers are in the back of the book.

1. Which, if any, of the following statements (a), (b), or (c), concerning a low-frequency shelf filter with a set of gain-vs.-frequency curves such as that shown in Fig. 11-11, is false?
   (a) It can be adjusted to function as a lowpass filter.
   (b) It can be adjusted to function as a bandpass filter.
   (c) It can be adjusted to function as a highpass filter.
   (d) All of the above statements (a), (b), and (c) are true.

2. Variable negative feedback in a closed-loop op amp circuit is employed to
   (a) maximize the gain.
   (b) minimize the gain.
   (c) control the gain.
   (d) produce oscillation.

3. Which of the following devices, assuming all the characteristics can be adjusted at will, is capable of producing a peaked response?
   (a) A rolloff filter.
   (b) A shelf filter.
   (c) A band-rejection filter.
   (d) A parametric equalizer.

4. Suppose an audio bandpass filter is provided with 8.8 mW of audio input power. At desired frequencies, the output power is 4.4 W. At unwanted frequencies, the output power is 2.2 mW. What is the ultimate power attenuation?
   (a) 27 dB.
   (b) 33 dB.
   (c) 54 dB.
   (d) 66 dB.
5. What is the insertion power gain of the filter described in the previous question at frequencies well within its passband?
   (a) 27 dB.
   (b) 33 dB.
   (c) 54 dB.
   (d) 66 dB.
6. An op amp circuit that amplifies signals outside of a defined frequency band, but attenuates signals within that band, is sometimes called
   (a) a shelf filter.
   (b) a parametric equalizer.
   (c) a closed-loop filter.
   (d) None of the above
7. A theoretically perfect op amp would have an input that appears as
   (a) a short circuit at all frequencies.
   (b) a short circuit at low frequencies and an open circuit at high frequencies.
   (c) an open circuit at low frequencies and a short circuit at high frequencies.
   (d) an open circuit at all frequencies.
8. Which of the following devices, assuming all the characteristics can be adjusted at will, is capable of producing a notch?
   (a) A bandpass filter.
   (b) A shelf filter.
   (c) A rolloff filter.
   (d) A parametric equalizer.

9. In the open-loop configuration, an op amp
   (a) has infinite output impedance.
   (b) exhibits its largest possible amplification factor.
   (c) cannot amplify because there is no feedback.
   (d) acts as an oscillator.
10. In a shelf filter, the shelf frequency is usually defined as
    (a) the frequency at the center of the sloping part of the gain-vs.-frequency curve.
    (b) the frequency at the point in the gain-vs.-frequency curve where the gain begins to increase or decrease from unity.
    (c) the frequency at the point in the gain-vs.-frequency curve where the gain reaches its maximum.
    (d) the frequency at the point in the gain-vs.-frequency curve where the gain reaches its minimum.

CHAPTER 12

Noise, Hum, Interference, and Grounding

Even when a hi-fi audio system is distortion-free and all components seem to be operating properly under lab conditions, trouble can still occur. Various electrical and electromagnetic (EM) effects, sometimes originating outside the system and beyond the control of the user, can cause havoc.

Internal Noise and Hum

Internal noise is produced by discrete components in electronic equipment, such as transistors, diodes, resistors, and integrated circuits (ICs). Minimizing the generation of this noise is a major priority in the design of hi-fi audio systems, particularly in circuits or devices that are followed by multiple stages of amplification.

THERMAL NOISE

Thermal noise occurs as a result of the constant, random motion of atoms and molecules. The level of thermal noise in any substance is proportional to the *absolute temperature*, which is measured relative to the absence of all heat (*absolute zero*, also known as *zero kelvin* or 0 K). Thermal noise decreases to near zero as the temperature approaches 0 K. In the everyday environment, however, temperatures are always well above 200 K. The freezing point of water, for example, is approximately 273 K. For this reason, thermal noise arises in all electrical and electronic media in the real world.

Thermal noise imposes a limit on the sensitivity that can be obtained with any electronic circuit or device. This noise can be almost totally eliminated by placing circuits in a bath of liquid helium. This element has a boiling point of only a few kelvins. Such cold temperatures cause the atoms and electrons to move slowly, reducing the thermal noise compared with the level at room temperature.

Of course, it isn't practical for the average audiophile to use *cryogenics* (such as a liquid-helium bath) in audio systems. This radical solution is pretty much confined to high-level scientific laboratories. But perhaps someday cryogenic audio systems will be available to consumers willing to pay a premium price for the ultimate in low-noise performance!

SHOT NOISE

In any current-carrying medium, the charge carriers (electrons and holes) cause noise impulses as they move from atom to atom. This is known as *shot effect*, and the resulting noise is called *shot noise*. All the shot noise in a circuit or device is amplified by succeeding stages along with the desired signals.

The amount of shot noise that a device produces is roughly proportional to the current that it carries. Low-current solid-state devices such as the *gallium-arsenide field-effect transistor* (GaAsFET) generate a minimum of shot noise. Devices of this sort, used in the most sensitive parts of an audio system, can help to ensure that the output at the speakers or headset contains the lowest possible level of amplified shot noise.

CONDUCTED NOISE

Electrical noise that enters an audio system through the power supply is called *conducted noise*. In a home audio system, this noise can be suppressed by inserting

radio-frequency (RF) *chokes* of about 1 millihenry (1 mH) in series with all the wires leading to the utility outlet(s). In addition, if necessary, *bypass capacitors* rated at 500 volts (500 V) or more, and having values of approximately 0.1 microfarad (0.1 μF), can be installed between each power wire and an *electrical ground*, representing a point of zero potential. In a properly wired electrical system, the "third slot" in each outlet is connected to the electrical ground for the building. Figure 12-1 illustrates the use of RF chokes and bypass capacitors for suppressing conducted noise.

In a vehicle, conducted noise comes from the alternator and spark plugs, through the DC power leads, and into the audio equipment. Conducted noise of this type can be minimized by connecting the power leads directly to the battery terminals, *not* to the cigarette lighter socket or to the socket for any peripheral electrical device such as an interior lamp. Filtering the non-grounded (usually positive) DC power lead using a series RF choke and a parallel capacitor, in an arrangement similar to that used for AC utility conducted noise, can minimize alternator whine. *Resistance wiring* in the ignition system sometimes helps to reduce conducted noise generated by the spark plugs in an internal combustion engine.

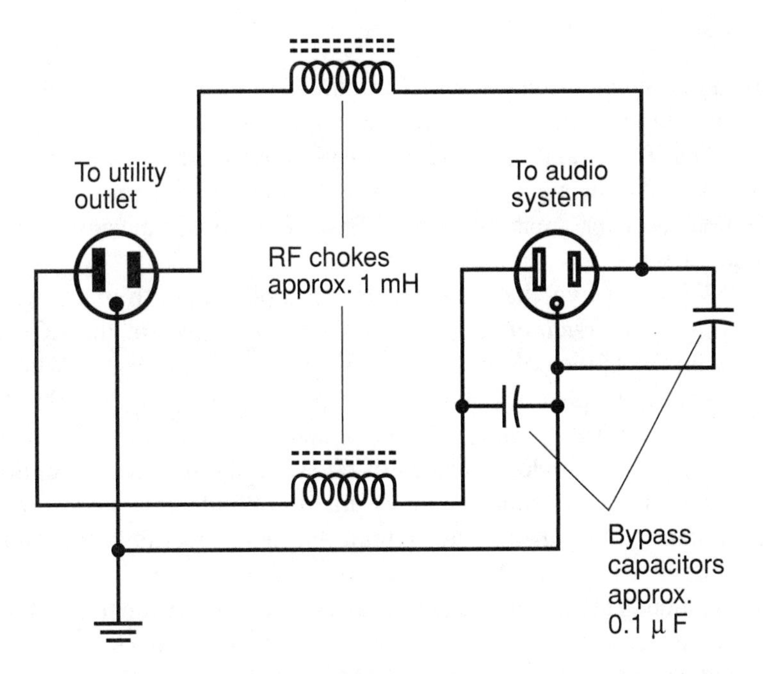

Figure 12-1 A line filter that can help to suppress conducted noise in an AC-operated home audio system.

MICROPHONICS

Mechanical vibration can cause unwanted modulation of the oscillators in a hi-fi tuner and also in the circuitry of an audio amplifier. Such unwanted modulation is known as *microphonics*.

In home audio systems, microphonics rarely poses a problem. Occasionally, trouble can occur if tuners, amplifiers, or peripherals are placed directly on top of speakers. In such cases, vibration of the speaker cabinets can be transmitted to sensitive circuits.

In vehicle audio systems, microphonics occurs more often. For maximum immunity to microphonics, internal circuit boards must be firmly anchored to the chassis, particularly in amplifiers. Component leads must be kept short, and connectors must be pushed or screwed firmly in place. Sensitive circuits such as oscillators may have to be encased in wax, foam, or some other shock-absorbing substance. It may be necessary to mount the entire audio electronic system (except for the speakers) with acoustic padding between the equipment cabinets and the vehicle dashboard.

RIPPLE

Ripple is the presence of undesired AC modulation of a signal or power source. The most common form is 60-Hz or 120-Hz modulation originating from AC power supplies. The output of any AC power supply contains some ripple, and its extent depends on how much current the supply is required to provide. The higher the current demand, the greater the ripple. Excessive ripple can cause audible *hum* in hi-fi equipment.

Figure 12-2 shows hypothetical examples of ripple output from a power supply that uses a *full-wave rectifier* (that operates on both halves of the AC cycle) as compared with the ripple output from a power supply that uses a *half-wave rectifier* (that operates on only one half of the AC cycle). Assuming a 60-Hz AC input, a full-wave rectifier has an *output ripple frequency* of 120 Hz, while a half-wave rectifier has an output ripple frequency of 60 Hz. In both of these examples, the *ripple filters* in the power supplies are assumed to be identical and to offer marginal performance. With a full-wave circuit, the hum level is lower than with a half-wave circuit when all other parameters are identical. For this reason, full-wave rectification is the *de facto* standard for power supplies intended for use with hi-fi audio equipment. Half-wave rectifiers are used mainly in low-current devices where a moderate amount of ripple can be tolerated.

Ripple filters consist of large-value inductors (on the order of several henrys) in series with the output of a power supply, and/or large-value *electrolytic capacitors*

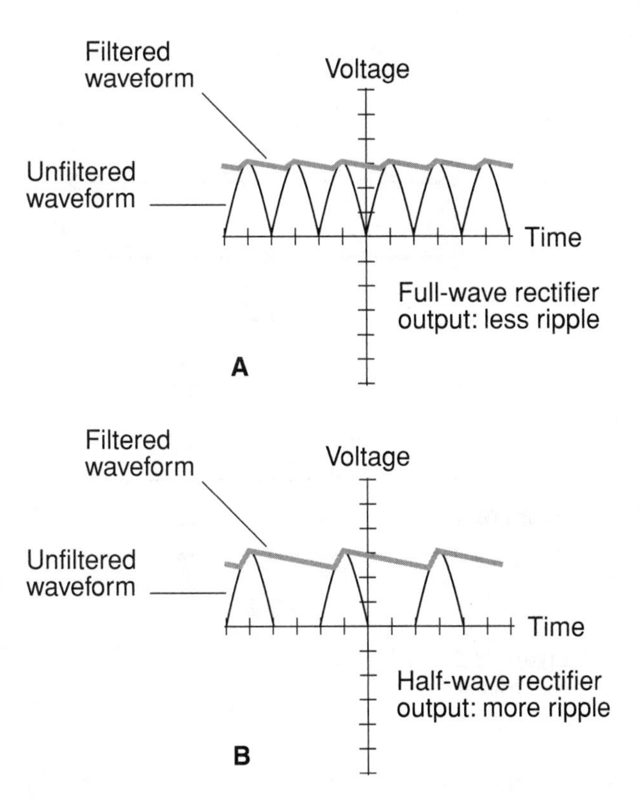

Figure 12-2 Hypothetical examples of ripple output from a full-wave rectifier (A) and from a half-wave rectifier (B).

(upwards of 100 μF) in parallel with the supply output. The more current the supply is required to deliver, the more inductance and/or capacitance is required in the ripple-filter stage. Figure 12-3 illustrates two effective ripple-filter configurations: a *capacitor-input ripple filter* (shown at A) and a *choke-input ripple filter* (at B). Note that electrolytic capacitors are *polarized.* The positive terminal, usually indicated by a plus sign (+) on the component, must be connected to the positive power-supply output terminal. Otherwise, the component will not function properly and may rupture or explode. Multiple ripple-filter stages of the same type (either choke-input or capacitor-input) can be cascaded to obtain enhanced performance.

Over-engineering of power supplies is another means by which ripple can be kept to a minimum. If a power supply can deliver a lot more current than will ever be demanded from it, then its filter circuits will never be "put to the test," and the filter components will be able to function at their best. Of course there is a practical,

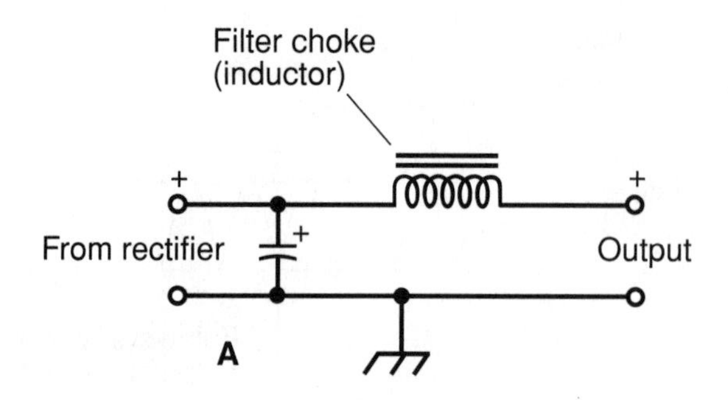

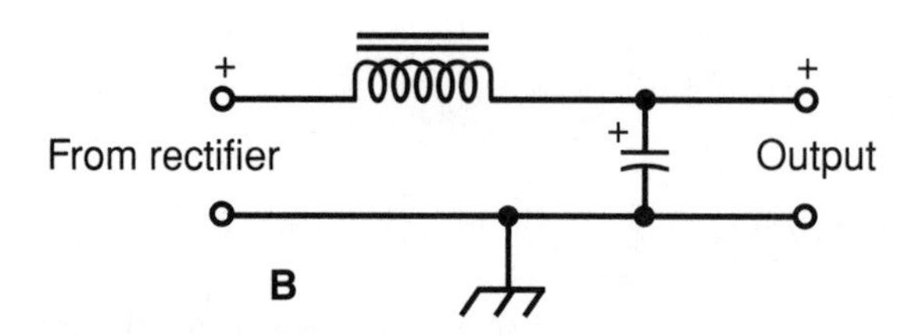

Figure 12-3 At A, a capacitor-input ripple filter. At B, a choke-input ripple filter.

as well as an esthetic, limit to this type of over-engineering. The use of a 10-amp (10-A) power supply with an amplifier that never draws more than 5 A constitutes a solid but reasonable configuration, but the use of a 50-A power supply with the same amplifier is a waste of money.

PROBLEM 12-1

Why does the ripple output from a power supply tend to increase as the current demand increases, if all other factors are held constant?

SOLUTION 12-1

As the current demand from a power supply increases, the resistance across its output terminals goes down. Another way of saying this is that the *load* becomes heavier. This causes the ripple-filter capacitors and/or chokes to discharge more rapidly, so the voltage output becomes "bumpier." Figure 12-4 illustrates this effect. Drawing A shows the voltage output from a hypothetical full-wave power supply when the current demand is low (light load); drawing B shows an example of the voltage output from the same supply when the current demand is high (heavy load).

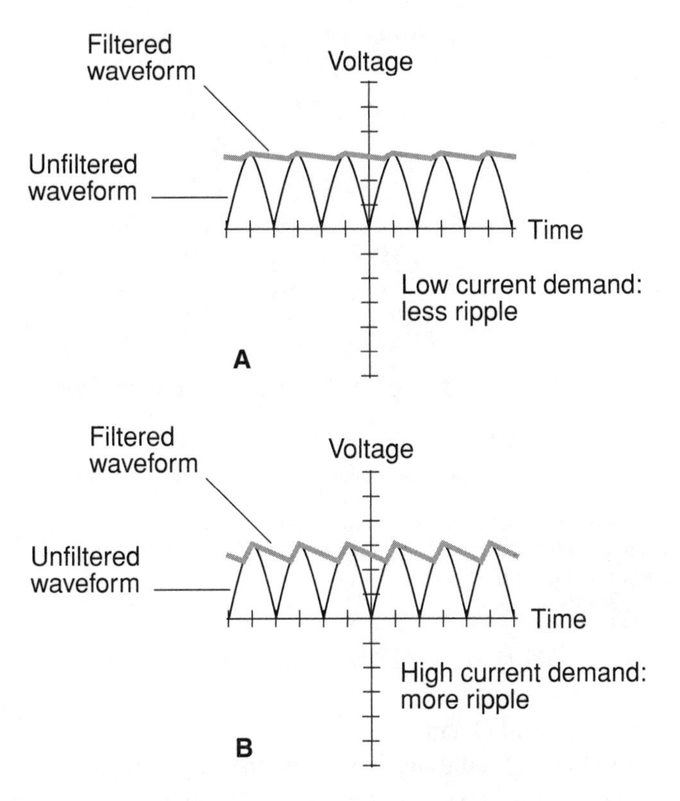

Figure 12-4 Illustration for Problem and Solution 12-1.

External Noise

External noise is mainly of concern in tuners, and consists of noise that originates outside a hi-fi sound system. Such noise can arise from natural or humanmade sources.

SFERICS

Electromagnetic (EM) *noise* is generated in the atmosphere, mostly by lightning. This noise is called *sferics*. In a radio receiver such as a hi-fi tuner, particularly in the amplitude (AM) broadcast band between 535 kHz and 1605 kHz, sferics produce a faint background hiss or roar, punctuated by sharp noise peaks often called "static." Figure 12-5 shows an example of sferics as they would appear on an

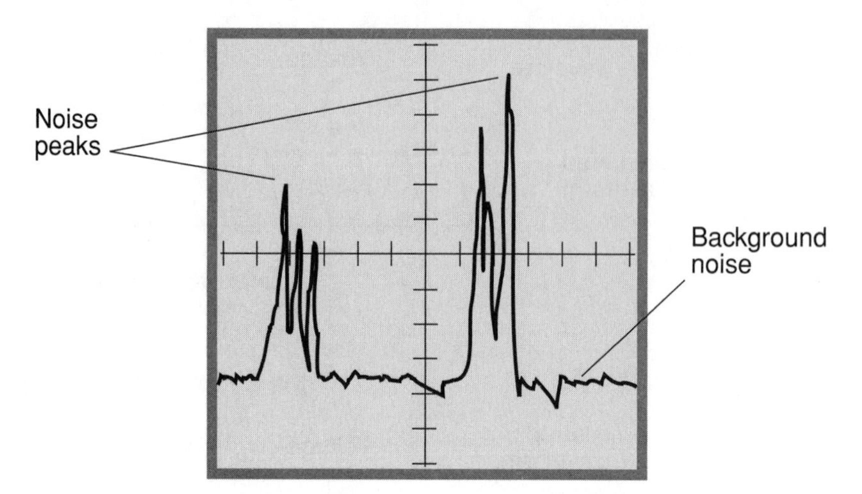

Figure 12-5 Sferics as they would appear on an oscilloscope connected into a hi-fi tuner system. Time is shown on the horizontal axis. Instantaneous amplitude is shown on the vertical axis.

oscilloscope connected to the intermediate-frequency (IF) section of a hi-fi tuner that is set to a vacant channel.

An individual lightning discharge, or *stroke*, produces a burst of EM energy at all radio frequencies as well as at infrared (IR), visible, and ultraviolet (UV) wavelengths. At low radio frequencies, sferics can propagate for hundreds of miles. As the frequency increases, the propagation distance decreases. That is one reason why sferics are more of a problem in the AM broadcast band than in the frequency-modulation (FM) band on a hi-fi tuner, which covers the range 88 MHz through 108 MHz. The other reason sferics are worse on AM is the fact that sferics are characterized by rapid fluctuations in amplitude, and FM receivers are designed to be relatively immune to such changes. Nevertheless, during a heavy thunderstorm, sferics can be heard on FM tuners, even in the presence of strong broadcast signals.

PRECIPITATION STATIC

Precipitation static is generated by electrically charged water droplets or ice crystals as they strike objects, particularly outdoor antennas that are not coated with insulating material. This type of interference is most likely to happen in snowstorms. Precipitation static can be severe, and can sometimes sound like a buzz or whine, as if it is coming from an electrical appliance or internal combustion engine.

In communications equipment, a common electronic method of dealing with precipitation static is the use of a *noise blanker*, which mutes the receiver during brief noise spikes. Most stereo FM tuners don't have noise blankers because they create brief silences, on the order of a few milliseconds, in the received sound. These silences can be tolerated in two-way radio communications systems where interference is part of the normal routine, but in a hi-fi sound system they can be more objectionable than the noise itself. Connecting a large-value inductor between the antenna and an electrically grounded point sometimes reduces or eliminates precipitation static, but not always.

In stereo hi-fi systems, the most effective overall way to deal with precipitation static is to use an indoor antenna for the tuner so rain, ice, and snow cannot reach the conductors. In remote areas far from strong broadcast stations, good performance with an indoor antenna can be difficult to obtain. However, locating the antenna as high up in the house as possible, such as in the attic, is the next best thing to an outdoor antenna as long as the house is not made of concrete and steel and doesn't have a metal roof.

CORONA

When the voltage on an electrical conductor exceeds a certain value, the air around the conductor ionizes. The result is *corona*, also called *Saint Elmo's fire*. This ionization causes broadband radio noise that can completely obliterate all received programs. Corona tends to occur in and near heavy thunderstorms. A pointed object, such as the end of a whip antenna, is more likely to produce corona than a flat or blunt surface.

It is a good idea to disconnect and ground all outdoor antennas during thunderstorms, because a nearby strike can induce an *EM pulse*, which is a surge of current sufficiently strong to damage or even destroy sensitive electronic equipment connected to the antenna. The existence of corona during a thunderstorm suggests the imminent possibility of a near or direct lightning strike.

IMPULSE NOISE

Any sudden, high-amplitude voltage pulse generates an electromagnetic (EM) field. In a radio receiver such as a hi-fi tuner, the result is *impulse noise*. This type of noise can be produced by household appliances such as vacuum cleaners, hair dryers, electric blankets, thermostat mechanisms, light dimmers, and fluorescent-light starters. It is also generated by internal-combustion engines. Impulse noise is, like sferics, more severe in the AM broadcast band than in the FM band, but it can often be observed in FM receivers when the signals are not strong enough to cause the limiter to effectively prevent amplitude fluctuations. Figure 12-6 is an example of impulse noise as it might appear on an oscilloscope connected in the IF chain of a tuner set to a vacant channel.

In sensitive hi-fi systems, impulse noise can be picked up by components other than the tuner. As the number of external peripherals (such as CD and tape players, computer interfaces, microphones, speakers, and headsets) in an audio system increases, the susceptibility to this type of noise also increases. Wireless microphones and headsets are especially vulnerable. The susceptibility of an audio system to impulse noise can be reduced by the use of a good ground system. Grounding is discussed later in this chapter.

POWER-LINE NOISE

Utility lines, in addition to carrying the AC that they are intended to transmit, carry high-frequency components that produce *power-line noise* in the form of RF

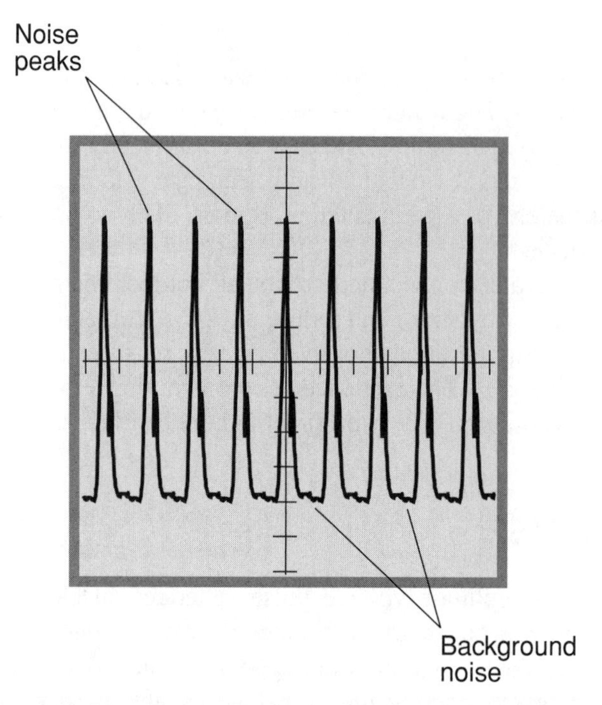

Figure 12-6 Impulse noise as it would appear on an oscilloscope connected into a hi-fi tuner system. Time is shown on the horizontal axis. Instantaneous amplitude is shown on the vertical axis.

fields. Power-line noise is a result of electric *arcing* (sparking) at some point in the circuit. The arcing can originate in appliances connected to terminating points. It can take place in faulty or obstructed power-line transformers. It can also occur in high-tension lines as a corona discharge into humid air. The RF produced by these sources is a widespread source of headaches for serious electronics hobbyists such as dedicated audiophiles and amateur radio operators.

Power-line noise sounds like a buzz, whine, or hiss when picked up by a hi-fi tuner. Some types of power-line noise can be attenuated in communications radios by means of noise blankers, but this is rarely a workable solution in hi-fi systems because of the momentary silences they produce. Special antennas with directional responses can sometimes provide dramatic relief from power-line noise. The antenna must be oriented so that a strong null in the response occurs in the direction from which the power-line noise is coming. The exact orientation must be determined by trial-and-error, and it usually varies as the frequency is changed. The null is quite sharp and can be difficult to find.

PROBLEM 12-2

Can AC utility-line current or voltage spikes be strong enough to upset the operation of hi-fi equipment or even damage it? If so, how can a system be protected against this threat?

SOLUTION 12-2

Power-line voltage spikes, technically called *transients*, are known to cause malfunctions in equipment that contains microcomputers. Most hi-fi equipment is microcomputer-controlled these days, and is therefore more sensitive to transients than was the case several decades ago. A good *transient suppressor*, also known as a "surge suppressor" or "surge protector," can offer some protection. But in order for such a device to work, it must be used with a properly installed three-wire AC electrical system. A nearby lightning strike can cause an extreme transient that can damage the power supply of a hi-fi system, even if a transient suppressor is used. The best protection for expensive equipment is to physically unplug it from the wall outlet during a thunderstorm.

Quantifying the Noise Level

There are several ways to define the extent of noise in relation to sound or signals in a hi-fi system. Noise can affect the overall quality and psychological effect of the sound, even when the noise isn't loud enough to be noticed by itself.

SIGNAL-TO-NOISE RATIO

The sensitivity of a radio receiver such as a hi-fi tuner is often specified in terms of the audio *signal-to-noise ratio* that results from an input signal of a certain number of microvolts. This ratio is symbolized as S/N or $S{:}N$.

Suppose the audio signal power at the speaker or headset terminals of a receiver is P_s, given in watts, and the audio noise level at the same speaker or headset terminals is P_n, also in watts. Then the signal-to-noise ratio, in decibels, is given by the following formula:

$$S/N = 10 \log (P_s/P_n)$$

Usually, the *S/N sensitivity* figure for a given receiver is specified as the input signal strength in microvolts rms, at the antenna terminals, that is required to produce $S/N = 10$ dB in the speaker or headset.

SIGNAL-PLUS-NOISE-TO-NOISE RATIO

The *signal-plus-noise-to-noise ratio* of a tuner is symbolized as $(S + N)/N$ or $(S + N){:}N$. This ratio is, like S/N, is always specified in decibels. Some manufacturers specify $(S + N)/N$ rather than S/N when quoting the sensitivity of a tuner.

If the audio signal power at the speaker or headset terminals of a receiver is P_s, expressed in watts, and the audio noise level at the same speaker or headset terminals is P_n, also in watts, then the signal-plus-noise-to-noise ratio, in decibels, is given by the following formula:

$$(S + N)/N = 10 \log [(P_s + P_n)/P_n]$$

Usually, the *(S + N)/N sensitivity* figure for a given receiver is defined as the input signal strength in microvolts rms, at the antenna terminals, that is required to produce $(S + N)/N = 10$ dB in the speaker or headset.

NOISE FIGURE

Noise figure (*NF*) is a specification that is sometimes used to evaluate the performance of a tuner or a weak-signal amplifier. Noise figure, like S/N or $(S + N)/N$, is expressed in decibels, and is a quantitative indicator of how much noise a circuit generates internally.

Noise figure is based on the amount of noise generated in a so-called *ideal circuit*. There is a difference between an ideal circuit and a *perfect circuit*. In a

perfect circuit, which can exist only in theory, there would be no noise generated whatsoever. But noise is an inevitable product of the motion of subatomic particles. In an ideal circuit there is always a little noise, but it is minimized to the greatest extent that contemporary engineering practice allows.

Let S_i be the *S*/*N* ratio (as defined above) in the speakers or headset of an ideal tuner or amplifier, and let S_r be the *S*/*N* ratio in the speakers or headset of a real-life system under test. The value of *NF*, in decibels, is calculated in terms of S_i and S_r according to this formula:

$$NF = 10 \log (S_i/S_r)$$

An ideal circuit exhibits $NF = 0$ dB. Noise-figure values of a few decibels are typical in practical tuners and amplifiers. Certain types of devices and circuits generate less noise than other types. For example, the GaAsFET is known for its low internal noise generation. This device can be useful in weak-signal amplifiers that are followed by several more stages of audio amplification. In low-level audio amplifiers, *NF* takes on exaggerated importance because the noise generated in the first stage of amplification is itself amplified by the subsequent stages, no matter how well-designed those subsequent stages are.

SINAD

The term SINAD is an acronym for the words *SIgnal to Noise And Distortion*. This expression can be used to comprehensively define the performance of an FM receiver under marginal conditions. The SINAD figure takes into account not only the sensitivity of a receiver, but its ability to reproduce a weak signal with a minimum of distortion.

Usually, the SINAD sensitivity of a receiver is given as the signal strength in microvolts rms at the antenna terminals that results in a SINAD ratio 12 dB in the speakers. The signal is modulated with a sine-wave tone at a specific frequency and deviation. The SINAD sensitivity is measured by using a calibrated signal generator and a specialized distortion analyzer called a *SINAD meter*. This instrument contains a notch filter centered at a standard audio frequency such as 1000 Hz. For determination of the SINAD sensitivity of a receiver, the procedure is as follows:

- The meter is connected to the speaker terminals of the receiver.
- The volume set near the middle of the control range.
- The meter-level control is set for a meter reading of 0 dB. Some SINAD meters have an *automatic level control* (ALC) that sets the meter at 0 dB regardless of the volume-control setting of the receiver.

- A signal generator is connected to the tuner antenna terminals, and the generator is set to the same frequency or channel as the tuner is set.
- The signal is modulated at the standard frequency and deviation.
- The signal level is adjusted until the SINAD meter indicates 12 dB. The SINAD sensitivity, in microvolts rms, is then read from the calibrated scale of the signal-generator amplitude control.

PROBLEM 12-3

Suppose a signal modulated with a 1000-Hz sine-wave tone is applied to the antenna terminals of a hi-fi tuner, and it produces 7.35 W in the speakers. Now imagine that the input signal is removed, so only internally generated noise appears at the speaker terminals. If this noise level is 354 mW, what is S/N for this receiver in this situation? What is $(S + N)/N$?

SOLUTION 12-3

First, convert both power figures to the same units. Let's use watts. Then in the above formula for S/N, we have $P_s = 7.35$ and $P_n = 0.354$. Plugging in these numbers:

$$\begin{aligned} S/N &= 10 \log (P_s/P_n) \\ &= 10 \log (7.35/0.354) \\ &= 13.2 \text{ dB} \end{aligned}$$

To determine $(S+N)/N$, plug the numbers into that formula, as follows:

$$\begin{aligned} (S + N)/N &= 10 \log [(P_s + P_n)/P_n] \\ &= 10 \log [(7.35 + 0.354)/0.354] \\ &= 10 \log (7.704/0.354) \\ &= 13.4 \text{ dB} \end{aligned}$$

Electromagnetic Interference

The term *electromagnetic interference* (EMI) refers to unwanted phenomena in which appliances, circuits, devices, and systems upset each other's operation because of EM fields they produce or pick up. When EMI is the result of an electronic system's improper response to nearby RF transmitters, it is sometimes called *radio-frequency interference* (RFI).

EMI FROM COMPUTERS

A personal computer (PC) produces wideband EM energy, especially if it has a cathode-ray tube (CRT) monitor. The digital pulses in the *central processing unit* (CPU) can also cause problems in some cases. Both of these sources of EM energy can be picked up by hi-fi tuners in the immediate vicinity. The energy gets out of the computer through the interconnecting cables and power cords, because they act as miniature transmitting antennas (Fig. 12-7A).

When a hi-fi system and a computer are placed in close proximity, such as when you want to amplify streaming audio from the Internet or want to have all your electronic systems in one place, you should not be surprised if EMI occurs when the tuner is set to certain frequencies. Some Internet connection technologies have also been known to cause EMI under certain conditions.

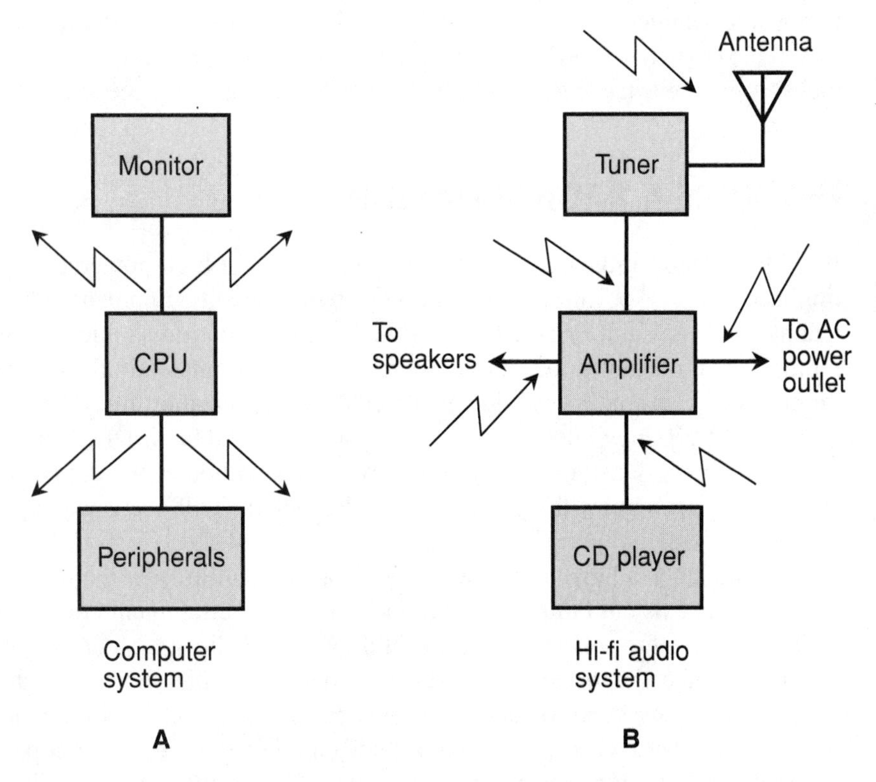

Figure 12-7 At A, a personal computer emits RF energy in its immediate vicinity. At B, a sound system can improperly pick up RF from nearby radio transmitters.

RFI FROM RADIO AND TELEVISION TRANSMITTERS

Hi-fi sound equipment can malfunction because of strong RF fields from a nearby radio or television broadcast transmitter. This can happen even when the transmitter is working exactly according to its specifications and in compliance with the standards set by the Federal Communications Commission (FCC). In these cases, and also in cases involving Citizens Band (CB) radios and amateur ("ham") radios, the root cause of the problem is rarely a transmitter malfunction. The trouble is usually the result of inferior home-entertainment-equipment design. The EM energy can get in through speaker wires, power cords, the tuner antenna, and cables between an amplifier and externals such as a CD player or tape deck. Some common vulnerabilities are shown in Fig. 12-7B.

As the number of connecting cables in a home entertainment system increases, the likelihood of interference from an RF field of a given intensity and frequency also increases. In addition, the risk of RFI increases as interconnecting cables are made longer. It is good engineering practice to use as few connecting cables in a home-entertainment system as possible, and to keep them as short as possible. If there is excess cable and you don't want to cut it shorter, coil it up and tape it in place. A good electrical ground for the entire system is also necessary.

IF YOU'RE A RADIO AMATEUR

If you're a "ham" radio operator with a sophisticated or high-powered station, you might be blamed for interference to home entertainment equipment, whether it is technically your fault or not. In situations of this kind, you should use the minimum amount of transmitter output power necessary to maintain the desired communication. Transmitters should be aligned to ensure that they are radiating signals only at the frequency intended. Antenna systems should be installed so as to radiate as little energy as possible into nearby homes and other buildings. Measures such as these minimize the chance that RFI problems will occur when you operate your station.

Unfortunately, a large proportion of RFI cases are ultimately the result of inadequate or nonexistent built-in protection for home entertainment equipment. This makes it difficult to solve the problem at the "ham" radio station. You may nevertheless mitigate the trouble by reducing your transmitter power, switching to another frequency band, or operating only when the home entertainment equipment is not in use. A compromising attitude can help to secure the cooperation of a neighbor who is experiencing RFI from your "ham" radio station.

It is unwise for a radio amateur to make modifications to a neighbor's home-entertainment equipment. If something goes wrong with it later, the "ham" could

be blamed. Once in a while, a manufacturer of home-entertainment equipment will offer technical support in RFI cases.

EMI FROM APPLIANCES AND POWER LINES

Tuners can pick up EMI from appliances such as vacuum cleaners, light dimmers, heating pads, electric blankets, hair dryers, and television sets. All of these devices contain components that generate electric sparks and/or produce harmonics of the AC utility wave because of nonlinear operation. Utility lines can also radiate considerable EM energy. Fortunately, these fields are rarely strong enough to interfere with consumer electronic systems directly, although they can cause trouble for shortwave radio listeners and amateur radio operators.

Power-line interference is caused by arcing between points that are in close proximity and differ greatly in voltage. A malfunctioning transformer, a bad street light, or a defective insulator can be responsible. This type of EMI can be difficult to locate and eradicate. Sometimes, help can be obtained by calling the utility company. If you happen to live near an amateur radio operator and you suspect that power-line interference is causing trouble with your hi-fi system, chances are good that the "ham" is also having trouble with it. The radio amateur's technical expertise may help you track down the source of the power-line noise if the utility company won't cooperate.

The internal combustion engines used in lawn mowers, weed trimmers, snow blowers, cars, trucks, farm implements, and road construction equipment occasionally cause EMI to hi-fi systems. This type of interference is similar to power-line EMI, but usually constitutes a less severe problem because of its intermittent or infrequent nature. In most situations of this kind, the offending device is easy to locate.

OTHER POTENTIAL PROBLEMS

Unwanted *RF mixing* can occur in the most unsuspected places. (This type of mixing, also called *heterodyning*, is not the same thing as the process that takes place in an audio mixing console.) At RF, *mixing products* are signals that arise at the sum and difference frequencies of other signals when they combine in a nonlinear device. Heterodyning can give rise to a species of RFI that affects radio receivers and hi-fi tuners, and that can be exceedingly difficult to track down and correct. Poor electrical connections in house wiring, plumbing, and exterior metallic structures such as fences and rain gutters can generate mixing products and harmonics in the presence of RF fields from multiple radio transmitters in the vicinity.

Intermodulation, a particularly obnoxious form of RF mixing, sometimes occurs in the downtown areas of large cities, where many radio transmitters are in operation simultaneously. The number of mixing products and harmonics in these areas can be so great as to constitute *broadband RF noise*. Intermodulation causes false signals in radio receivers, often sounding "hashy" or broken-up. In the worst cases, FM stereo reception can be devastated by intermodulation.

Interference to hi-fi tuners can sometimes be caused by harmonics or by *spurious emissions* (output signals at frequencies other than the design frequency) from a nearby broadcast, CB, or amateur radio transmitter. This does not occur very often with well-designed CB and amateur-radio transmitters because they employ relatively low power. But if you live in the shadow of a big broadcast or cellular communications tower, you might experience trouble.

PREVENTIVE MEASURES

Various tactics can be employed in the quest to keep stray RF energy out of home entertainment equipment. One such scheme is the same as that used to minimize conducted noise, described earlier in this chapter. Multiple RF chokes and bypass capacitors can be installed in power cords and interconnecting cables. You must be sure that these components won't interfere with the transmission of power, signals, or data through cables. For advice, consult the manufacturer of the equipment in question, or a competent engineer. If you try to install RF chokes or bypass capacitors yourself, they might not work properly. You also run the risk of voiding the equipment warranty if the installation involves internal modification or the cutting of built-in cords or wires.

RF shielding can help to prevent sensitive electronic apparatus from picking up stray RF fields. The simplest way to provide RF shielding for a circuit or device is to surround it with metal, usually copper or aluminum, and to connect this metal to a good electrical ground. Because metals are good conductors, an RF field sets up electric currents in them. These currents produce a secondary RF field that opposes the original field. Therefore, if the metal enclosure is well grounded, the RF field is effectively shorted out. Obviously you aren't going to wrap all your hi-fi gear in aluminum foil; but if you're shopping for a new system, particularly a tuner or amplifier, a metal cabinet is a feature to look for.

In addition to the use of metal cabinets, interconnecting cables must be shielded for optimum protection against RFI. In a shielded cable, all the signal-carrying conductors are surrounded by a tubular copper braid that is electrically grounded through the connectors at the ends of the cable. The most popular form of shielded cable is *coaxial cable*. It can often replace two-wire cords between an amplifier and other parts of a system.

As more and more RF-generating consumer devices are built, manufacturers ought to feel increasing pressure to provide RFI protection in electronic equipment. The *American Radio Relay League* (ARRL), 225 Main Street, Newington, CT, publishes books about EMI and RFI phenomena, their causes, and ways to deal with problems when they occur. These publications are intended mainly for amateur radio operators, but savvy hi-fi users may find them useful as well.

PROBLEM 12-4

Give an example of how the harmonic emissions from a radio transmitter could affect a nearby hi-fi FM radio receiver.

SOLUTION 12-4

Imagine that there is a transmitter a few hundred meters from your house that produces a powerful signal at 48.95 MHz. This is the *fundamental frequency* of the transmitter. Let's call it f_1. Then the *harmonic frequencies* (call them f_2, f_3, f_4, and so on) are as follows:

$$f_2 = 48.95 \text{ MHz} \times 2 = 97.9 \text{ MHz}$$
$$f_3 = 48.95 \text{ MHz} \times 3 = 146.85 \text{ MHz}$$
$$f_4 = 48.95 \text{ MHz} \times 4 = 195.80 \text{ Mhz}$$

and so on. Now suppose you want to listen to a weak station at 97.9 on the dial of your hi-fi tuner. Unless the transmitter has exceptional *harmonic suppression*, you can expect to get interference from the *second harmonic* of the transmitter, or the signal at f_2 = 97.9 MHz.

Grounding

All electronic equipment must be connected to a good electrical ground system. Grounding is important for personal safety. It can help protect equipment from damage if lightning strikes nearby. Grounding minimizes the chances for EMI and RFI problems with the equipment, and ensures that RF shielding will be maximally effective.

GROUNDING FOR SAFETY

There's an unwritten commandment in electrical and electronics engineering: "Never touch two grounds at the same time." This refers to the exasperating tendency for a high voltage to exist between different system points when you don't suspect it, and even when you think it is impossible. Many people have received severe electrical shocks, and more than a few have died, because they forgot this rule. It is an excellent idea to wear insulating gloves and rubber-soled shoes when working on electrical systems, even if you think such measures constitute "overkill."

Most new electronic devices have three-wire cords. One wire is connected to the "common" or "ground" part of the hardware, and leads to a D-shaped or U-shaped prong in the plug. You can recognize three-wire electrical systems by the appearance of the wall outlets. If each outlet has a sideways-D-shaped hole below the two rectangular holes, then you have a three-wire system. But for the system to be effective, the contact in this D-shaped hole must actually be connected to an earth ground. Sometimes, three-wire outlets are installed in a home or business, but the grounding system is not hooked up; the ground points in the outlets are actually left "floating"! This is worse than a two-wire system because it can deceive people into thinking they have electrical ground protection when in fact they do not.

GROUNDING FOR TRANSIENT SUPPRESSION

A grounded three-wire electrical system is necessary if transient suppressors are to work properly. All good transient suppressors have three-prong outlets. Never defeat the third (ground) prong on a transient suppressor. If you use an extension cord with your equipment, be sure it is the three-wire type, keep it as short as possible, and place the transient suppressor at the equipment location, not at the outlet. Also—and this can't be overemphasized—*be certain that the ground prongs in the outlets are connected to an actual ground!* In an older home with new additions, the electrical system in the original rooms may have three-wire outlets installed for cosmetic reasons, but employ the old two-conductor electrical wiring inside the walls. The outlets in the new additions might be all right. But don't count on that. It is wise to have an electrician check them out and correct any problems if they are found. This will not only maximize the likelihood that the measures you take against EMI and RFI will work, but it will also help to keep your equipment from getting "zapped"!

MINIMIZING EMI AND RFI

A good *RF ground* can help minimize EMI and RFI to home entertainment equipment. Figure 12-8 shows a proper RF ground scheme (A) and an improper one (B). In a good RF ground system, each device is connected to a single *ground bus*,

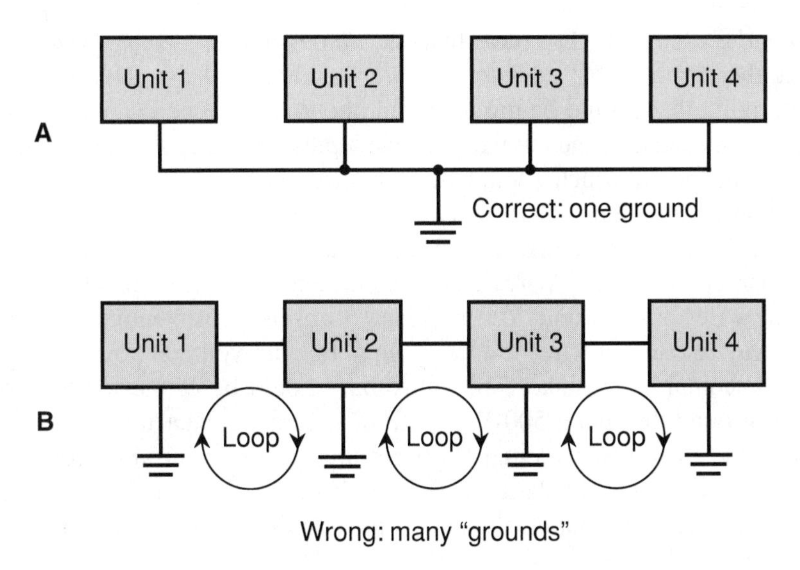

Figure 12-8 At A, the correct method for grounding multiple units in an audio hi-fi system. At B, an incorrect method that creates RF ground loops. Illustration for Problem and Solution 12-5.

which in turn runs to the earth ground through a single wire. This wire should be of a heavy gauge (AWG No. 12 is good), and as short as possible. A poor ground system contains *ground loops* that can act like RF antennas. If one of these loops happens to *resonate* at the frequency of a potentially interfering RF signal, the risk of RFI is especially high.

PROBLEM 12-5

All the units in my hi-fi system have plastic cabinets, and all but one of them has a two-wire electrical cord. How can I ground this system to minimize the chance that I will experience EMI or RFI problems?

SOLUTION 12-5

This isn't easy, and unless you actually experience RFI, you might as well not bother with it. But if you are pretty sure you are having trouble with EMI or RFI, you can take the following steps.

First, *be sure all your hi-fi equipment is unplugged from the wall outlets.*

Check the back of each unit for ground-connection lugs. If any of the units have them, connect each one of them individually to a bus as shown in Fig. 12-8. For units that do not have ground lugs, certain external terminals are connected to zero-potential, or *common*, points in the internal circuitry. Such a point is at *effec-*

tive ground for the unit. On a tuner that has a provision for a coaxial (unbalanced) antenna, the terminal that goes to the outer conductor of the cable represents a common point that should be grounded. In phono and phone plugs and jacks for microphones, headsets, and auxiliary device inputs, the outer sleeves are normally at a common point, which should be grounded. Figures 12-9 and 12-10 show phono plugs (also called *male connectors*, at A in each illustration) and jacks (also called *female connectors*, at B in each illustration). Note the conductors and terminals marked "Ground." Wires can be connected from these points to the bus, using one wire for each unit. You will have to use some ingenuity to figure out the best way to connect wires to these points. Be sure you don't ground a signal-carrying terminal by mistake! The entire bus should be connected to a known electrical ground through a 500-V, 0.1-μF disk-ceramic capacitor. This will pass most RFI but prevent accidental short circuits in the event your power outlet happens to have reversed polarity. If you aren't sure where to find a "known electrical ground," consult your electrician. Copper cold-water pipes are often, but not always, grounded.

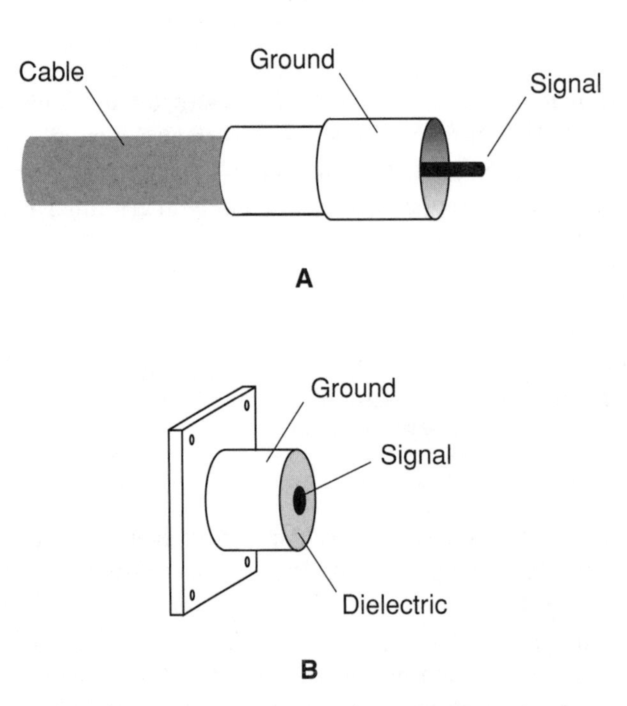

Figure 12-9 A phono plug (A) and jack (B). Illustration for Problem and Solution 12-5.

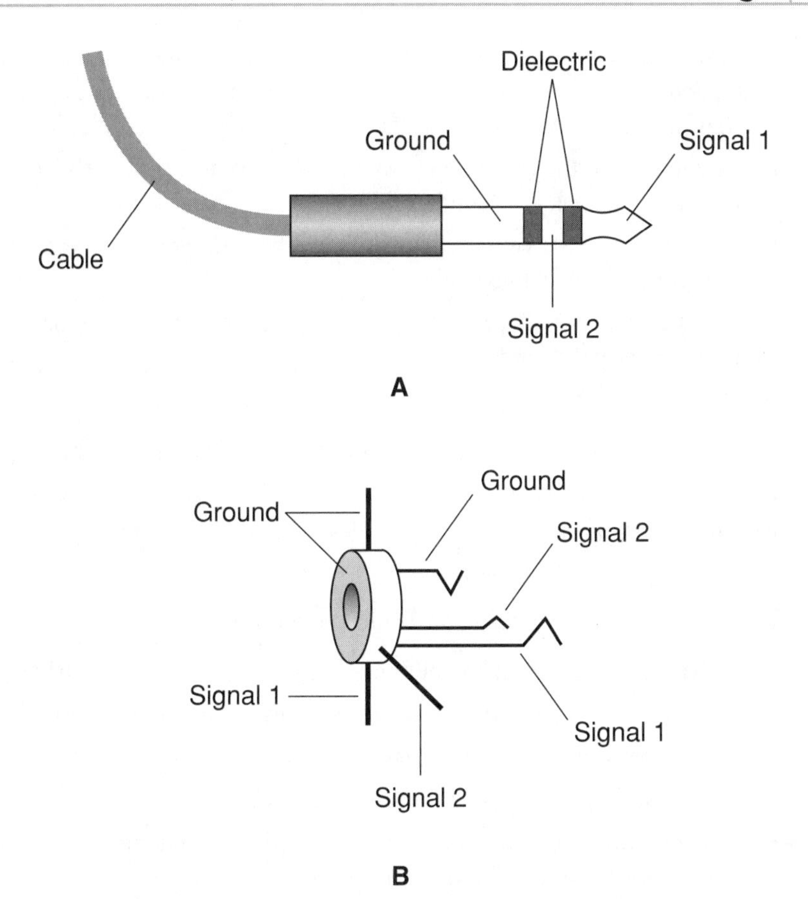

Figure 12-10 A stereo phone plug (A) and jack (B). Illustration for Problem and Solution 12-5.

Only after all the preceding steps have been completed should you plug your equipment back into the wall outlets. And remember again the time-worn adage, "Never touch two grounds at the same time."

Quiz

This is an "open book" quiz. You may refer to the text in this chapter. A good score is 8 correct. Answers are in the back of the book.

1. Suppose you have bought a new house, and you see three-conductor outlets in the room where you plan to put your hi-fi system. You haven't had an electrician inspect the house. Which of the following statements is true for certain?
   (a) The longer of the two vertical slots in each outlet is directly connected to a good electrical ground.
   (b) The shorter of the two vertical slots in each outlet is directly connected to a good electrical ground.
   (c) The sideways-D-shaped hole in each outlet is directly connected to a good electrical ground.
   (d) None of the above
2. The EMI produced by a malfunctioning utility power transformer
   (a) results in harmonic emissions.
   (b) causes heterodyning.
   (c) can be difficult to locate.
   (d) does not affect amateur or CB radio operators.
3. Excessive shot noise, when generated in a microphone preamplifier,
   (a) can cause subsequent stages to malfunction, and may even damage them.
   (b) is amplified in subsequent stages along with the desired audio.
   (c) can cause improper operation of the microphone.
   (d) can be eliminated by the use of impedance-matching transformers between subsequent stages of amplification.
4. What is the best way to minimize problems with precipitation static in a hi-fi stereo system that uses an FM tuner?
   (a) Sharpen the tips of all the antenna elements.
   (b) Locate the antenna indoors.
   (c) Install a surge suppressor between the tuner and the wall outlet.
   (d) Install a filter capacitor across the antenna terminals.
5. In a real-life tuner or amplifier that generates some noise, $(S + N)/N$ in the presence of an input signal
   (a) is greater than S/N.
   (b) is less than S/N.
   (c) is equal to S/N.
   (d) may be greater than, less than, or equal to S/N, depending on the circumstances.

6. Which of the following measures (a), (b), or (c) can help to minimize the chance that you will have problems with conducted noise in an automotive hi-fi system?
   (a) Connect the hi-fi system power leads directly to the battery terminals.
   (b) Connect a capacitor in series with the positive power-supply lead.
   (c) Connect a capacitor in series with the negative power-supply lead.
   (d) Any of the above measures (a), (b), and (c) can help to minimize the chance that you will have problems with conducted noise.

7. Suppose an FM tuner is set for 99.9 MHz. An interfering signal is heard, and you suspect that it is a harmonic from a radio transmitter somewhere in the vicinity. If this is in fact the case, on which of the following output frequencies might the offending transmitter be operating?
   (a) 299.7 MHz.
   (b) 199.8 MHz.
   (c) 33.3 MHz.
   (d) 22.2 MHz.

8. Thermal noise can be almost completely eliminated by
   (a) using bipolar transistors in amplifier stages.
   (b) using resistances that are as low as possible.
   (c) avoiding nonlinear operation of transistors.
   (d) immersing all system components in liquid helium.

9. Full-wave rectifier circuits are standard in audio equipment, in part because
   (a) they can produce high voltages.
   (b) they produce relatively little ripple.
   (c) they are ideal for low-current applications.
   (d) All of the above

10. Suppose a signal modulated with a sine-wave audio tone is applied to the antenna terminals of a hi-fi tuner, and it produces 8.535 W in the speakers. Now imagine that the input signal is removed, so only internally generated noise appears at the speaker terminals. If this noise level is 8.535 mW, what is *S*/*N* for this receiver in this situation?
    (a) 1000 dB.
    (b) 60 dB.
    (c) 30 dB.
    (d) It cannot be determined without more information.

CHAPTER 13

Recording, Reproduction, and Synthesis

This chapter discusses how sound is recorded, reproduced, processed, and synthesized in hi-fi systems. You'll also learn how speech can be generated and interpreted electronically.

Analog Audio Tape

Analog tape recording is the process of impressing audio signals onto magnetic tape so the amplitude and frequency can be varied over continuous ranges. *Analog tape playback* is the process of recovering the signals and obtaining the original sound waveforms. An *analog audio tape recorder* can perform both of these functions.

THE PHYSICAL MEDIUM

Magnetic tape, also called *recording tape*, has countless ferromagnetic particles attached to a flexible, thin plastic strip. In the analog audio tape recorder, a fluctuating magnetic field produced by the *recording head* polarizes these particles. As the field changes in strength, the tape passes by at constant speed. This creates regions on the tape in which the ferromagnetic particles are polarized to various extents in either direction. A hypothetical input waveform is illustrated in Fig. 13-1A. The resulting polarization pattern is graphed at B, where N represents magnetic north and S represents magnetic south on the vertical axis.

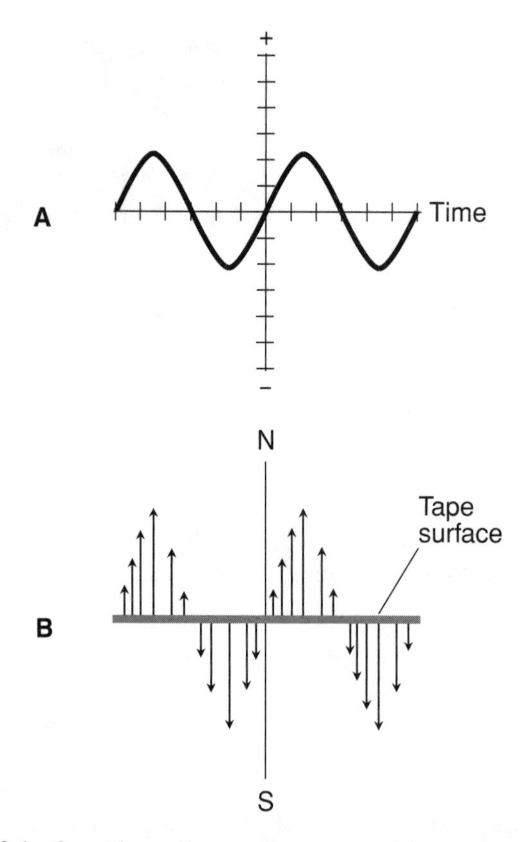

Figure 13-1 On analog audio recording tape, particles are magnetized in a pattern that follows the input waveform. Graph A shows an example of an analog audio input waveform. Graph B shows relative polarity and intensity of magnetization for selected particles on the tape surface.

When the tape is run through the recorder in the playback mode, the magnetic fields around the ferromagnetic particles cause a fluctuating magnetic field that is detected by the *pickup head*. This field has the same pattern of variations as the original field from the recording head.

Analog audio tape is available in various widths and thicknesses. Thick tapes don't play as long for a given mass (total quantity or weight) as thin tape, but thick tape is more resistant to stretching. The impulses on a magnetic tape can be distorted or erased by external magnetic fields. Therefore, magnetic tape should be protected from such fields. Extreme heat can also result in loss of data, and can cause permanent physical damage to the tape by warping or melting it.

FEED SYSTEMS

A typical audio tape *cassette* plays for 30 minutes on each side; longer playing cassettes allow recording for as much as 60 minutes per side. Four recording paths, each with two channels for the left and the right half of the sound track, provide a total of eight recording tracks. To use a cassette, you simply open up a small door on the tape recorder, snap the cassette into place, and close the door. Some cassette tape recorders have a slot into which you can slide the cassette. To remove a cassette, you press an "eject" button that causes a mechanical lever to dislodge the cassette.

A *reel-to-reel tape feed system* resembles that of an old-fashioned movie projector. The tape is wound on two flat spools called the *supply reel* and the *take-up reel*. The reels rotate as the tape passes through the recording/playback mechanism. When the take-up reel is full and the supply reel is empty, both reels can be flipped over and interchanged. This allows recording or playback on the "other side" of the tape. (Actually, the process takes place on the same side of the tape, but on different tracks.) A reel 7 in (18 cm) in diameter can hold about 3000 ft (914 m) of 0.5-mil-thick tape. The speed is usually 1.875, 3.75, 7.5, or 15 in/s. This corresponds to 4.76, 9.53, 19.1, and 38.1 cm/s, respectively. Slower speeds are suitable for voice recordings, but the higher speeds provide better sound reproduction for music and special effects.

OPERATION

Analog audio tape recorders act as transducers between acoustic signals and variable magnetic fields. Figure 13-2 is a simplified rendition of the recording and playback apparatus in a typical analog audio tape recorder. The same basic scheme is used in cassette machines and reel-to-reel machines.

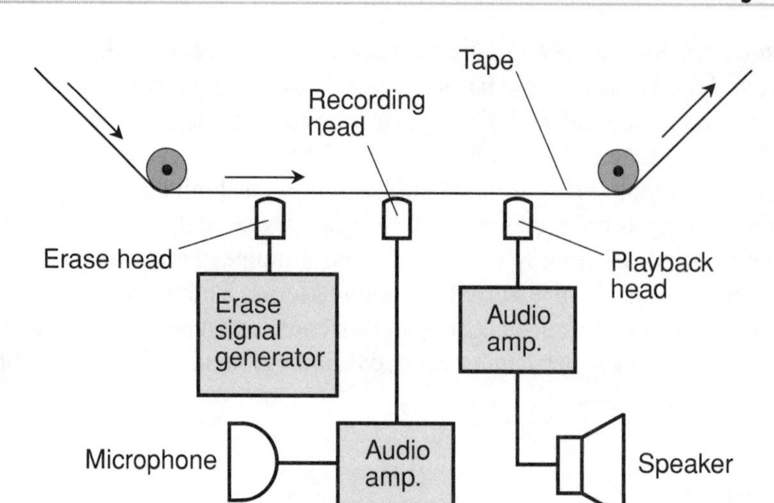

Figure 13-2 Simplified functional diagram of the recording and playback hardware in an analog audio tape recorder.

In the *record mode*, the tape moves past the *erase head* before anything is recorded on it. If the tape is not blank (that is, if magnetic impulses already exist on it), the erase head removes these impulses before anything else is recorded. This prevents *doubling*, or the simultaneous presence of two programs on the tape. The erase head can be disabled in some tape recorders when doubling is desired. The *recording head* is a miniature, precision electromagnet that generates a fluctuating magnetic field whose instantaneous flux density is proportional to the instantaneous level of the audio input signal. This magnetizes the tape in a pattern that duplicates the waveform of the signal. The *playback head* is usually not activated in the record mode. However, the playback head and circuits can be used while recording to create an *echo* effect.

In the *playback mode*, the erase head and recording head are not activated. The playback head acts as a sensitive magnetic-field detector. As the tape moves past the playback head, the head is exposed to a fluctuating magnetic field whose waveform is identical to that produced by the recording head when the audio was originally recorded on the tape. This magnetic field induces weak alternating currents in the playback head. The currents are amplified and delivered to a speaker, headset, or other output device.

PROBLEM 13-1

Why are higher tape speeds preferred for music and special effects? Isn't all sound the same, encompassing a frequency range from 20 Hz to 20 kHz?

SOLUTION 13-1

The human voice contains sounds concentrated mostly below 3 kHz. For effective reproduction of a voice where fidelity is not of concern, an analog tape recording system only needs to be sensitive to frequencies up to that limit. But music and special effects contain significant sound components as high as 20 kHz (and even beyond, into the ultrasonic range). The ability of analog audio tape to record high-frequency sound increases in direct proportion to the speed at which the tape travels past the recording head. This is because there is a limit to how closely together the magnetized regions can be "bunched" on the tape. The higher the tape speed, the higher the maximum audio frequency that can be efficiently recorded for a given spacing of the magnetized regions.

Digital Audio Tape

Digital audio tape (DAT) is magnetic recording tape on which *binary digital data* can be recorded. Digital recording provides better fidelity, greater *dynamic range* (the difference between the maximum and minimum loudness without objectionable distortion, expressed in decibels), and a better noise figure than analog recording.

DIGITAL VERSUS ANALOG SIGNALS

A signal is said to be *digital* when it can attain only a finite number of discrete levels or states, usually some power of two. This is in contrast to analog quantities that vary continuously, and can theoretically attain an infinite number of levels or states. The sounds you hear in everyday life are analog in nature. But they can all be converted to digital impulses consisting of *binary digits* (called *bits*). The bits are grouped into larger units called *bytes*.

When analog sound is converted into digital form, it can be more effectively recorded than can the original analog signal. But in order for a listener to appreciate a digital recording, the digital impulses must be converted back into the original analog signal before it is sent to the speakers or headset.

ANALOG-TO-DIGITAL CONVERSION

Any analog voice or music signal can be *digitized*, or converted into a string of impulses whose amplitudes can have only certain levels. This process is known as *analog-to-digital* (A/D) *conversion.* A circuit called an *A/D converter* or ADC parses the analog signal into "slices" and then rounds their amplitudes off to specific whole-number values. The number of possible states, or levels, of each slice is equal to some power of two, so it can be represented as a binary-number code. The number of possible digital states is called the *sampling resolution*, or simply the *resolution*. As the resolution increases, so does the precision with which the audio signal can be digitized. A resolution of 2^3, or 8 (as shown in Fig. 13-3), is standard for commercial digital voice circuits. A resolution of 2^4, or 16, is good enough for most hi-fi sound systems that record or reproduce music.

The accuracy with which a signal can be digitized depends on the frequency at which the sampling is done. In general, the *sampling frequency*, also called the *sampling rate*, in A/D conversion must be at least twice the highest frequency component in the original analog signal. For a voice signal, the commercial standard is 8 kHz (8000 samples per second). For music and hi-fi digital transmission, the standard sampling rate is 44.1 kHz (44,100 samples per second).

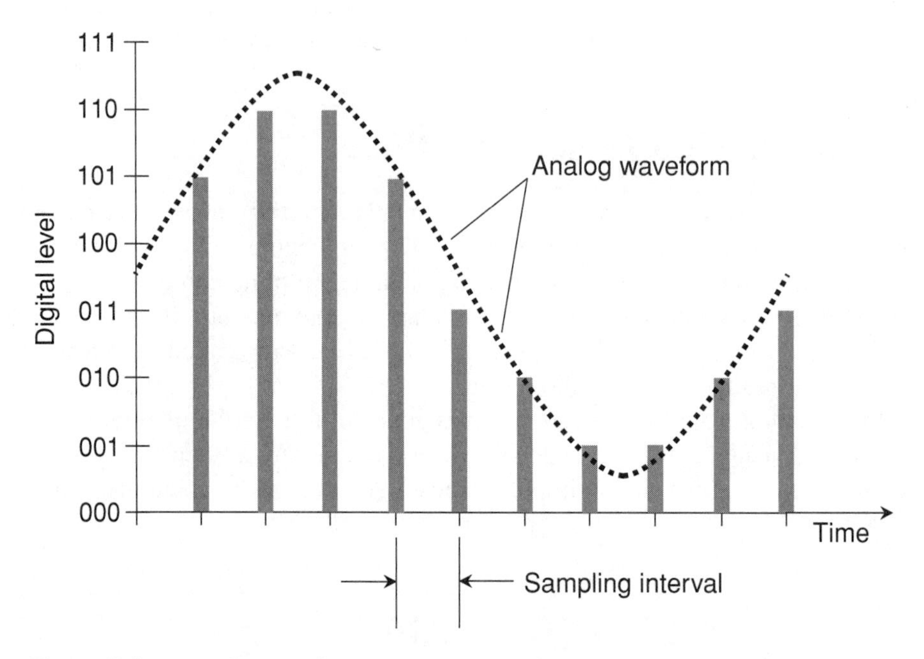

Figure 13-3 An analog waveform (dashed curve) and an 8-level digital representation (vertical bars).

DIGITAL SIGNAL PROCESSING

Digital signal processing (DSP) is an electronic method of improving the precision of digital data. Digital signal processing was first used with radio and slow-scan television receivers to produce a clear voice and/or picture from a marginal signal. Some of the earliest experimenters with DSP were amateur radio operators.

Digital signal processing can improve the fidelity and reduce the noise in digitized audio. It is easier for a machine to process a digital signal with its finite number of well-defined states, than to process an analog signal which has a theoretically infinite number of possible states. In particular, digital signals have specific patterns that are easy for microprocessors to recognize and clarify, whereas in an analog signal, the patterns are infinitely varied and complicated.

The DSP circuit works by getting rid of confusion between digital states. The circuit makes a "decision" between the high and low state for defined time intervals. If the incoming signal is above a certain level for an interval of time, the DSP circuit output is high (logic 1). If the level is below the critical point for a time interval, then the circuit output is low (logic 0).

DIGITAL-TO-ANALOG CONVERSION

The reverse of A/D conversion is *digital-to-analog* (D/A) *conversion*. This process allows the original analog data to be recovered from a digitized audio signal. You can use Fig. 13-3 to envision D/A conversion. Think "in reverse" from the way you imagine A/D conversion. A circuit called a *D/A converter* or DAC levels out the digital impulses, so the original analog signal emerges.

ADVANTAGES OF DAT

Noise is reduced in digital recording because noise is analog in nature, and is practically ignored by digital circuits. Some electronic noise is generated in the analog amplification stages following D/A conversion, but this is minimal compared with the noise generated in fully analog systems. The reduced noise and superior dynamic range in DAT equipment provides more true-to-life reproduction than is possible with analog methods.

Another advantage of DAT is the ability to make multi-generation copies. While analog signals are "fuzzy" in the sense that they vary continuously, digital signals are "crisp." Imperfections in the recording apparatus, the tape, and the pickup head affect digital signals to a lesser extent than they affect analog signals. Digital signal processing eliminates the flaws that would otherwise creep into a digital signal each time it is recorded and played back. The fidelity does not deteriorate to a noticeable extent even after many recording/reproduction copy cycles.

PROBLEM 13-2

Draw a waveform diagram showing how DSP can "clean up" a binary signal containing analog noise.

SOLUTION 13-2

Digital signal processing minimizes analog noise in a binary signal as shown in Fig. 13-4. A hypothetical signal before DSP is shown at A; the signal after processing is shown at B. If the incoming signal amplitude is above a defined level for an interval of time, the DSP output is logic 1. If the amplitude is below that level for a time interval, then the output is logic 0. By processing multiple binary digital signals in *parallel* (simultaneously), this technique can be applied to digital signals having a number of states equal to any power of two.

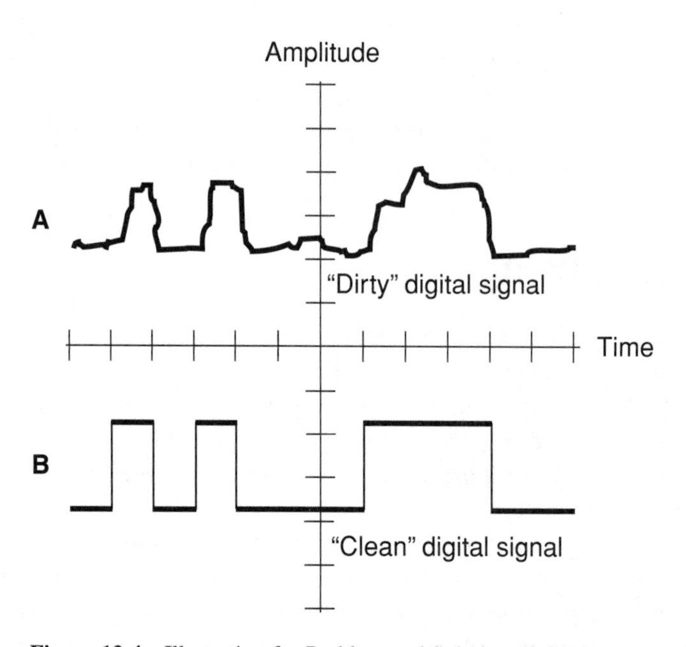

Figure 13-4 Illustration for Problem and Solution 13-2.

Compact Disc

A *compact disc* (CD) is a flat plastic disc with a diameter of 4.72 in (12 cm) on which data is recorded in digital form. Digitized audio on the surface of a CD is practically devoid of the hiss and crackle that historically bedeviled recordings on *vinyl discs*.

HOW IT WORKS

When a CD is manufactured, the analog audio is subjected to A/D conversion. The resulting digital bits are then burned by a laser into the surface of the disc. This forms microscopic pits in the plastic. The pits are arranged in a long spiral track that measures several kilometers in length.

When the sound is to be played back, another laser beam (much weaker than the one used to prepare the original disc) scans the CD without physically affecting the surface. The laser beam is scattered by the pits, and is reflected from the unpitted plastic. The result is a digitally modulated beam that is picked up by a sensor and converted into electrical currents. That digital signal is subjected to DSP to get rid of noise and to optimize the dynamic range. Finally the processed digital signal passes through a D/A converter, is amplified, and is transmitted to the speakers or headset.

COMPARISONS WITH VINYL AND TAPE

Compact-disc players recover the sound from a disc without physically touching the surface on which the data is recorded. Therefore, no matter how many times a CD is played, it will not suffer physical erosion. This is in contrast to the old vinyl discs that "wore out" after hundreds of passes of the *stylus* ("needle") through the groove.

If you want to change selections on a vinyl disc, you must lift the stylus from the surface and then place it at the beginning of the *band* of grooves representing the song you want to hear. If you want to move to a specific point within a selection, you have to make a guess as to where that point might be, and plunk the stylus down! If your hand slips or trembles, the stylus can scratch the vinyl.

With magnetic tape, selection of songs requires fast-forwarding and rewinding. It can take two or three minutes to reach the selection you want. Even if the tape player is programmed to locate selections automatically, the process takes time. It's unavoidable because of the physical geometry of the system.

With a CD player, all the above mentioned processes are electronic, and they can all be done fast. Selections are assigned numbers that you select by pressing buttons. It is impossible to damage the CD, no matter how much you skip around among the songs. You can "fast-forward" or "rewind" to different points within an individual selection. You can program a CD player to play only those selections you want, ignoring the others.

RECORDABLE AND REWRITABLE CDS

There is a type of CD on which you can write data, but once it has been written, it's there to stay. You cannot change it later. This is called a *compact disc recordable* (CD-R). There are other CDs that allow you to overwrite old data, just as you can with magnetic tape. This is called a *compact disc rewritable* (CD-RW).

Most personal computers now come with CD-R and/or CD-RW drives built in. The drives can record and play back CDs containing audio, video, multimedia, computer programs, images, text—anything that can be digitized. These drives, along with the transition to digital audio in general, have made the computer a standard component of high-end audio recording and reproduction systems. Gone are the days of physically cutting and splicing magnetic tape to create an audio production! (Of course, some people still prefer to work with tape and scissors, just as some people enjoy driving and tinkering with old cars.)

PROBLEM 13-3

How can data from analog audio tape be put on a digital CD?

SOLUTION 13-3

The analog tape is played back on a standard tape recorder, and the output of this device is fed directly to a computer with a CD-R or CD-RW drive. Computer programs are available to facilitate this process. It may be necessary to place *attenuator pads* (networks of resistors) between the left- and right-channel outputs of the tape recorder and the left- and right-channel audio inputs of the computer to prevent overdriving the computer sound board. The computer should have as much memory as possible, and the sound board should be of the highest possible quality. All you need to do once the electronic connections have been made is play the tape and run the computer program to put the audio file on the CD.

Speech Synthesis

Speech synthesis, also called *voice synthesis*, is the electronic generation of sounds that mimic the human voice. These sounds can be generated from digital text files or from printed documents. Speech can also be generated by computers, in the form of responses to stimuli or input from humans or other machines. One of the more interesting applications of this technology is its use as an alternative to the human voice in musical productions.

WHAT IS A VOICE?

A frequency band of 300 Hz to 3000 Hz is wide enough to convey all the information, and also all of the emotional content, in a human voice. In hi-fi applications, however, it is best if the system can generate sounds over the entire range of human hearing. This ensures good fidelity as well as intelligibility. The big challenge is to produce waves at exactly the right frequencies, at the right times, and in the right phase combinations. The modulation must also be correct, so the intended meaning is conveyed.

In the human voice, the volume and frequency rise and fall in subtle and precise ways. The slightest change in modulation can make a tremendous difference in the meaning of what is said. You can tell, even over the telephone, whether the speaker is anxious, angry, or relaxed. A request sounds different than a command. A question sounds different than a declarative statement, even if the words are the same.

SPEECH VARIABLES

In the English language there are 40 elementary sounds, known as *phonemes*. In some languages there are more phonemes than in English; some languages have fewer phonemes. The exact sound of a phoneme can vary, depending on what comes before and after it. These variations are called *allophones*. There are 128 allophones in English. These can be strung together in countless ways.

The *inflection*, or "tone of voice," is another variable in speech; it depends on whether the speaker is angry, sad, scared, happy, or indifferent, as well as on age, gender, and culture. A voice can also have an *accent*. You can tell when a person speaking to you is angry or happy, regardless of where that person hails from. Some accents sound more authoritative than others; some sound funny if you haven't been exposed to them before. Along with accent, the choice of word usage varies in different regions. This is *dialect*. A high-level speech synthesizer takes all of these variables into account.

RECORDING, PLAYBACK, AND READING

The most primitive form of speech synthesizer is a set of recordings of individual words. These can be put on a magnetic tape, a magnetic disc, a CD, a computer hard drive, or an electronic storage medium known as *flash memory*. You will sometimes hear the voices from such "speech synthesizers" in automatic telephone answering machines and services. They all have a characteristic choppy, interrupted sound. There are several drawbacks to these systems. Perhaps the biggest problem is the fact that each word requires a separate recording. If the recordings are on tape or disc, they must be mechanically accessed, and this takes time.

Printed text can be read by a machine using *optical character recognition* (OCR), and converted into a standard digital code called *ASCII* (pronounced "ASK-ee"). This is an acronym for *American Standard Code for Information Interchange*. The ASCII code can be translated by a D/A converter into voice sounds, so a machine can read text aloud.

Because there are only 128 allophones in the English language, a machine can be designed to read almost any text. But a machine lacks a sense of which inflections are best for the different scenarios that occur in a narrative. With technical or scientific text, this is rarely a problem. But in reading a fiction story to a child, mental imagery is crucial. No machine yet devised can paint pictures or elicit moods in a listener's mind as well as a human being.

THE PROCESS

There are several ways in which a machine can be programmed to produce speech. A simplified block diagram of one process is shown in Fig. 13-5. No matter what scheme is used for speech synthesis, certain steps are necessary. These are as follows:

- The machine must access the data and arrange it in the proper order.
- The phonemes and allophones must be assembled in the correct sequence.
- The proper inflections must be put in.
- Pauses must be inserted in the proper places.

In addition to the above, features such as the following can be included for additional versatility and realism:

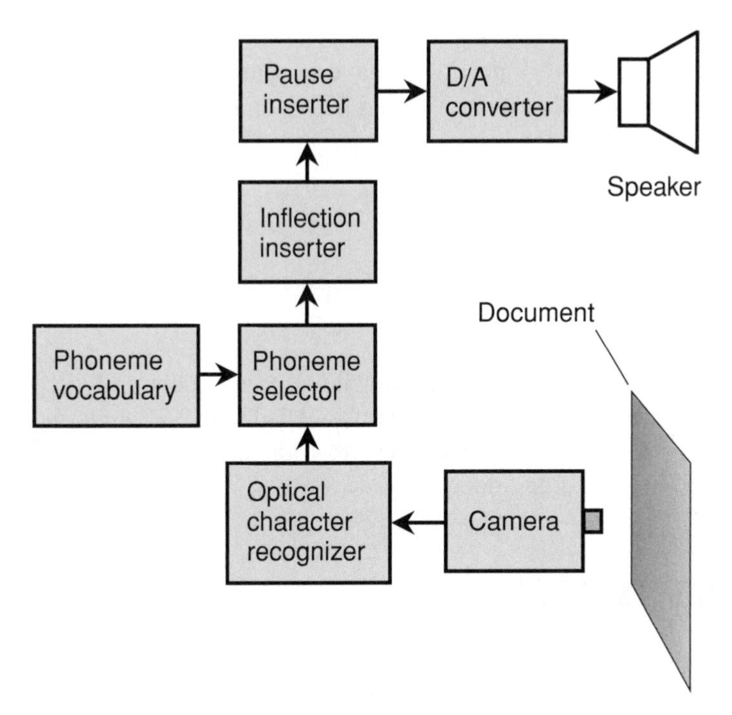

Figure 13-5 Block diagram of a speech-synthesis system that translates printed text into an intelligible voice.

- An intended mood can be conveyed (joy, sadness, urgency, etc.) at various moments.
- Overall knowledge of the content can be programmed in. For example, the machine can know the significance of a story, and the importance of each part within the story. This generally requires a *supercomputer* with *artificial intelligence* (AI).
- The machine can have an interrupt feature to allow conversation with a human being. If the human says something, the machine will stop and begin listening with a *speech-recognition* system. This, too, requires AI.

PROBLEM 13-4

Suppose you have written a tune that you think would make a good popular music production, but you don't have a musical group. There are four different tracks intended for singing voices. Can a speech synthesizer be used to generate all four of these tracks, so the whole production is, in effect, sung by a computer?

SOLUTION 13-4

Technically, this can be done if you have a computer with enough memory and processing power. However, it would be beyond the economic reach of most individuals.

Speech Recognition

Computers, and computer-controlled machines, can be controlled by means of spoken commands. *Speech recognition*, also called *voice recognition*, makes this possible. Figure 13-6 is a simplified block diagram of a speech-recognition system intended for control of a computer.

COMPONENTS OF SPEECH

Suppose you speak into a microphone that is connected to an oscilloscope, and see the jumble of waves on the screen. How can any computer be programmed to make sense out of that? The answer lies in the fact that, whatever you words you utter, it is made up of only a few dozen phonemes. These phonemes can be identified by computer programs.

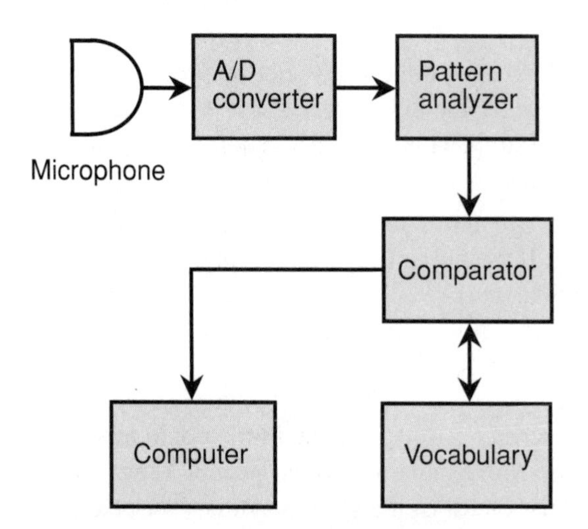

Figure 13-6 Block diagram of a speech-recognition system that can be used to input commands to a computer.

Most of the acoustic energy in a human voice occurs within three defined frequency ranges, called *formants*. The *first formant* is at less than 1000 Hz. The *second formant* ranges from approximately 1600 Hz to 2000 Hz. The *third formant* is at approximately 2600 Hz and up. Between the formants there are *spectral gaps*, or ranges of frequencies at which little or no sound occurs. A frequency-vs.-time graph of speech creates an image called a *voice print*.

The formants, and the gaps between them, remain within the same frequency ranges no matter what is said. The fine details of the voice print determine not only the words, but all the emotions, insinuations, and other aspects of speech. Any change in "tone of voice" shows up in a voice print. Therefore, in theory, it is possible to build a machine that can recognize and analyze speech as well as any human being. Such a machine can work fairly well if the whole range of input frequencies for the analog voice signal is restricted to the band from 300 Hz to 3000 Hz.

PULSE MODULATION

The A/D converter in a speech synthesis system changes the continuously variable (analog) voice signal into a series of digital pulses. Several different characteristics of a digital pulse train can be varied. These processes are collectively known as *pulse modulation* (PM). Four types of PM are briefly described here, and are diagrammed in Fig. 13-7 as amplitude-vs.-time graphs.

In *pulse amplitude modulation* (PAM), the strength of each individual pulse varies according to the instantaneous amplitude of the analog input waveform (Fig. 13-7A). The pulses all have equal duration and are equally spaced. *Pulse duration modulation* (PDM), also called *pulse width modulation* (PWM), is shown at B. In this method, the pulses all have equal amplitude and their time centers are equally spaced, but they last for different lengths of time, depending on the instantaneous amplitude of the analog input signal. *Pulse interval modulation* (PIM), also called *pulse frequency modulation* (PFM), is shown at C. In this scheme, the pulses all have equal amplitude and equal duration, but they occur more or less often depending on the instantaneous amplitude of the analog input signal.

In *pulse-code modulation* (PCM), any of the aforementioned aspects—amplitude, duration, or frequency—of a pulse train can be varied. But rather than having infinitely many possible states or levels, there are only a few. Usually this number is 8 or16. Figure 13-7D shows an example of 8-level PCM in which the pulse amplitude is varied. Pulse-code modulation is the technology of choice for speech recognition because the discrete signal levels can be defined by sequences of bits. For example, in an 8-level signal, three bits suffice, corresponding to binary numbers 000 (for level 0) through 111 (for level 7). In a 16-level signal, four bits suffice, corresponding to binary numbers 0000 (for level 0) through 1111 (for level 15).

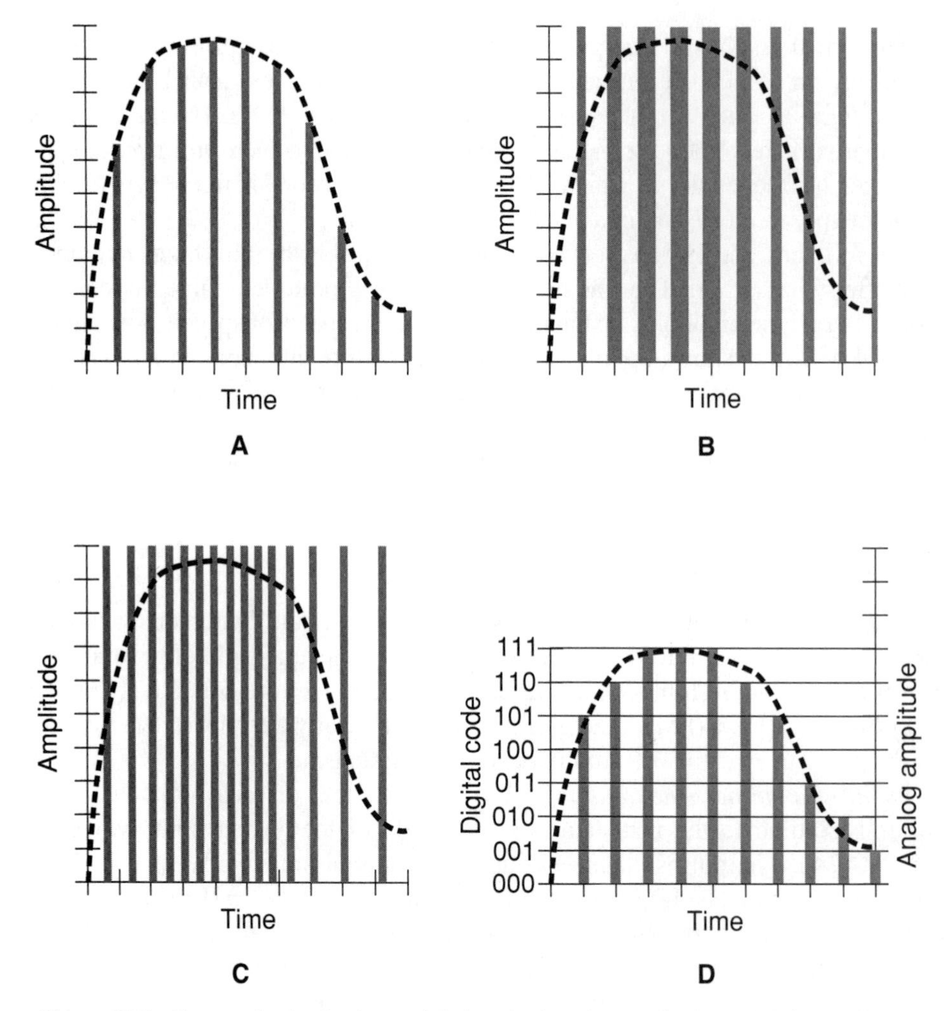

Figure 13-7 Four methods of pulse modulation. At A, pulse amplitude modulation; at B, pulse width modulation; at C, pulse interval modulation; at D, pulse code modulation. Dashed curves represent analog input waveforms. Vertical bars represent pulses.

A digital signal can convey a human voice within a passband about 200 Hz wide. That is less than 10 percent of the passband of the analog signal. The narrower the bandwidth, in general, the more the emotional content is sacrificed. But in speech recognition, in which the output device is a text display, printer, computer, or robot, emotional content is conveyed by the way in which the words are sequenced (that is, according to *context* and *syntax*), and not by "tone of voice."

WORD ANALYSIS

For a computer to decode a digital voice signal, it must be stocked with a *vocabulary database* of words or syllables, and some means of comparing this knowledge base with the incoming audio signals. This system has two parts: a *memory*, in which various speech patterns are stored, and a *comparator* that compares these stored patterns with the data coming in. For each syllable or word, the circuit checks through its vocabulary until a match is found. This is done very fast, so the delay (called *latency*) is not noticeable. The size of the computer real-time-usable vocabulary is directly related to its memory capacity and processing power. An advanced speech-recognition system requires a large amount of memory and a fast microprocessor. A big hard drive with plenty of free storage space doesn't hurt, either.

The output of the comparator must be processed in some way, so that the machine knows the difference between words or syllables that sound alike. Examples are "two/too," "way/weigh," "four/for," and "not/knot." For this to be possible, the context and syntax must be examined. The computer must be able to determine whether a group of syllables constitutes a part of a word, one word, two words, three words, or more. The more complicated the voice input, the greater the chance for confusion. Even the most advanced speech-recognition system makes mistakes, just as people sometimes misinterpret what you say.

INSINUATIONS AND EMOTIONS

The pulse-code modulation (PCM) process in a basic binary digital speech-recognition system removes the inflections from a voice. In the extreme, all of the tonal changes are lost, and the voice is reduced to "audio text." For most speech-to-text and machine-control purposes, this is adequate. If a system could render every word with the correct spelling and could insert all the punctuation properly, speech-recognition engineers would be pleased indeed! Eventually this ideal will be approached, and then interest will turn to rendering the subtler meanings of speech for advanced applications such as robot control.

Consider the sentence "You will visit Dan's grocery store after midnight," and say it with the emphasis on each word in turn (eight different ways), as follows:

- *You* will visit Dan's grocery store after midnight.
- You *will* visit Dan's grocery store after midnight.
- You will *visit* Dan's grocery store after midnight.
- You will visit *Dan's* grocery store after midnight.
- You will visit Dan's *grocery* store after midnight.

- You will visit Dan's grocery *store* after midnight.
- You will visit Dan's grocery store *after* midnight.
- You will visit Dan's grocery store after *midnight*.

The meaning changes dramatically depending on the *prosodic features* of your voice: which syllables or words you emphasize, and which ones you utter normally or de-emphasize. Prosodic features are important for another reason, too. A sentence might be a statement or a question. Thus "You will visit Dan's grocery store after *midnight*?" represents something completely different from "You will visit Dan's grocery store after *midnight*." Even if all the tones are the same, the meaning can vary depending on how fast something is said. The timing of breaths can make a difference, as well.

Prosodic features can be included in a PCM digital signal, but a larger number of bits are required than is necessary to render "audio text" without expression included. In a computer system this translates to (you guessed it!) the need for more memory and processing power.

Music Synthesis

Anyone interested in audio technology and music production has heard the term *synthesizer* (or "synth"). These devices have existed for decades. If you're an audio veteran, you'll remember the *Moog synthesizer* from the 1960s. Nowadays, music synthesis can be done in many ways, and the process can be controlled by a computer.

SOFTWARE SYNTHESIS

Suppose you connect a microphone through an amplifier to the vertical input of an oscilloscope, so you can observe the waveforms produced by sounds. You play middle C on a piano and observe the resulting waveform (Fig. 13-8). If you play middle C on a clarinet, you'll see a different waveform. A trombone will produce still another waveform.

Now imagine this process in reverse. Instead of playing an instrument and looking at the wave, you draw the wave with a mouse or electromechanical pointing device, using a computer program that converts whatever you draw into a periodic waveform. Then you feed the waveform to an amplifier and speaker. If you draw the waveform you see on an oscilloscope for a particular musical instrument, you'll get a note from the speaker that sounds as if it is played by that instrument. You can

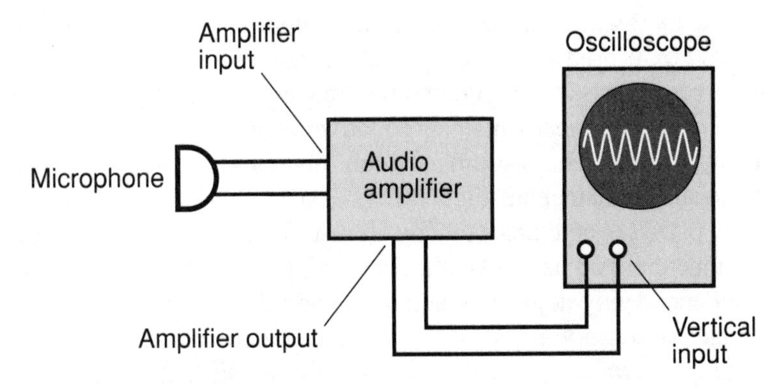

Figure 13-8 The waveform of a musical note can be viewed on an oscilloscope connected to a microphone and an audio amplifier.

also draw waveforms that do not resemble those of known musical instruments, thereby generating unique-sounding musical notes.

The above described way of generating electronic music constitutes *software synthesis*, one of the most advanced music synthesis technologies. The advent of the personal computer has made software synthesis possible for the home musician. But you don't have to own a computer to generate electronic music.

ANALOG SYNTHESIS

Analog synthesis is the oldest method of generating electronic music. It employs simple oscillators, amplifiers, and filters using transistors and op amps. Vacuum tubes are used in some designs dating back to the 1960s.

Analog synthesis is preferred by some musicians who claim the resulting sound is more rich and full than that produced by digital means. This may be in part because analog devices are a little imperfect by nature, and the real world is an imperfect place! The frequency of an analog oscillator has a tendency to change slightly with time (this is called *drift*).

The main limitation of analog synthesis is the fact that it is difficult to create sounds that have exactly the waveform you want. Special effects can't be programmed; you have to literally "play it by ear."

ADDITIVE SYNTHESIS

An electronic oscillator can be designed to produce a pure sine wave in which all the energy is concentrated at a single frequency, but real-life musical instruments

produce sound energy at multiple frequencies. Suppose you play an instrument at a pitch of 600 Hz. It produces waves at the harmonics (1200 Hz, 1800 Hz, 2400 Hz, and so on) as well as at the fundamental frequency. The amplitudes of the harmonics relative to the fundamental, and relative to each other, vary depending on the type of instrument. The instrument might also produce waves at frequencies not related to the fundamental. Figure 13-9 is a frequency-domain display of a hypothetical "600-Hz" note produced by an instrument with a complex output.

Now imagine that you have a synthesizer with dozens of sine-wave oscillators, each with independently adjustable amplitude and frequency controls. If you set the frequencies and amplitudes of these oscillators to produce the frequency-domain display shown in Fig. 13-9, you'll get a sound that mimics that of the hypothetical instrument playing a "600-Hz" note. The effects of all the different sine waves add up to "build" a musical note. This process is called *additive synthesis*. It, like software synthesis, is best done using a computer with plenty of memory and processing power, along with a high-end sound card.

SUBTRACTIVE SYNTHESIS

Another way to produce complex musical notes is to begin with a signal produced by a special oscillator called a *comb generator*. This oscillator produces a sine wave at a specific frequency, along with (in theory) all possible harmonics of that

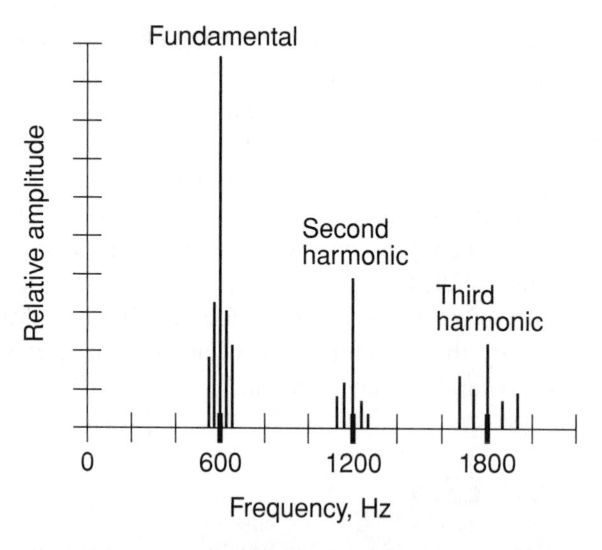

Figure 13-9 Frequency-domain display of a complex "600-Hz" note played by a hypothetical instrument.

frequency. In an ideal comb generator, the fundamental and the harmonics all have equal amplitudes. (The expression "comb" derives from the appearance of the pips on a frequency-domain display, resembling the teeth of a comb.) The output of the comb generator is passed through a high-end parametric equalizer. Unwanted sine-wave components are attenuated. The desired signals are allowed to pass with variable gain to produce a specific pattern such as that shown in Fig. 13-9. The end result of this process, called *subtractive synthesis*, is exactly the same as it would be if the note had been "built" by additive synthesis.

Subtractive synthesis is easier to perform than additive synthesis, and does not require the use of a computer. Think of additive synthesis as an attempt to create a statue by gluing together tiny bits of stone, and subtractive synthesis as the more familiar sculpture method, in which a large block of rock is carved away until only the desired material remains.

WHITE-NOISE SCULPTING

An alternative method of subtractive synthesis involves the tailoring of *white noise* to produce signals. Noise, in contrast to signals, is defined in terms of *spectral power density*, a measure of the amount of sound power contained within a narrow band of frequencies having a constant width such as 1 Hz. Spectral power density can be specified in units of watts, milliwatts, or microwatts per hertz (W/Hz, mW/Hz, or μW/Hz). White noise has the same spectral power density throughout the whole audio spectrum. Therefore, the amount of sound power between 550 Hz and 551 Hz is the same as that between 2450 Hz and 2451 Hz, or between 12.343 kHz and 12.344 kHz, or between 34.7 Hz and 35.7 Hz.

White-noise sculpting requires the use of a circuit called a *white-noise generator* in conjunction with a sophisticated parametric equalizer. Figure 13-10 is a hypothetical example of a "600-Hz" musical note produced in this fashion. Note that there is some energy at the second and third harmonics as well as at the fundamental frequency. But rather than consisting of signals that occur at sharply defined frequencies, the components of this note are "fuzzy." This method of synthesis produces characteristically "airy" notes. Sometimes these notes sound like those produced by a flute played softly. But more often, music produced in this fashion sounds unlike any real-life instrument.

SAMPLING

In the *sampling* method of music synthesis, small digital wave files are created for various instruments played at specific frequencies and intensities. These wave files can be stored on the hard drive of a personal computer, or on any other digital

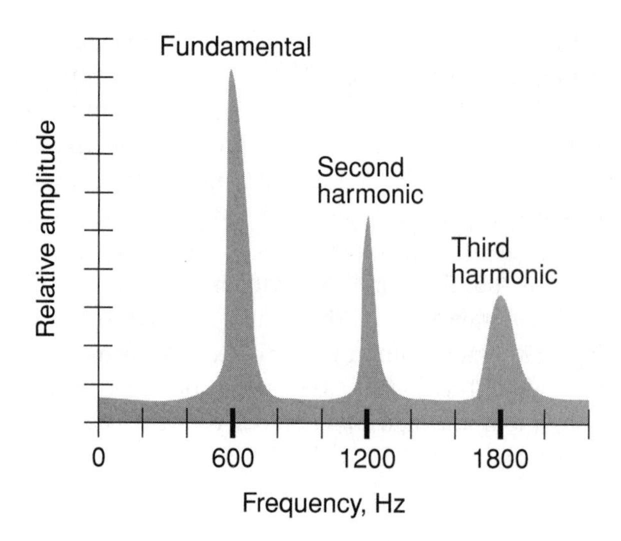

Figure 13-10 Frequency-domain display of a "600-Hz" note produced by sculpting white noise.

medium such as CD-R, CD-RW, external hard drive, flash memory, or tape. The files can also be exchanged over the Internet.

Suppose you record the notes for all the familiar band and orchestra instruments at fundamental frequencies of 30 Hz, 100 Hz, 300 Hz, 1 kHz, 3 kHz, and 10 kHz. You also record the sounds of drums and other percussion instruments. (In fact, you can record anything, even shouting children or barking dogs!) For each of these files, you record the notes or effects as the instruments are played at low, medium, and high intensity. Thus, you build a repertoire of several hundred wave files. These files can be processed by a computer program to generate notes of variable characteristics such as staccato, vibrato, and so on. Another computer program can combine notes into chords, and chords into compositions. With patience and practice, you can compose intricate and unique-sounding musical pieces in digital format. You might even create a 21st-century symphony this way. (One must wonder what would Mozart or Beethoven would think if they could witness all this!)

MUSICAL INSTRUMENT DIGITAL INTERFACE

Musical instrument digital interface, more commonly called by its acronym MIDI (pronounced "MID-ee"), is a programming language used in electronic music. A computer can be used with MIDI to compose, edit, store, transmit, or download

electronic notes and tunes. The MIDI program tells the computer when to play each note, how long to play it, and how loud to play it. It also sets the tempo of the music, based on how long a *quarter-note* lasts.

A MIDI program is something like the roll of paper that runs through an old-fashioned player piano, except that a MIDI program is electronic rather than mechanical, and it is much more versatile. The MIDI program can control the operation of multiple music synthesizers.

Standard components in a system equipped with MIDI include the computer, monitor, modem, printer, and auxiliary drives. In addition, there is a MIDI processing unit, a MIDI keyboard (that resembles an electronic organ), and at least one synthesizer, also called a *MIDI sound module*. Some MIDI sound modules can produce only a few waveforms, representing the more well-known musical instruments such as the piano, clarinet, and trumpet. Others include such exotic sounds as the bongo drums, balalaika, sitar, chirping birds, baying wolves, or peeping porpoises. Using a MIDI program with a good module is like playing a computer game, except that you have more control over the results.

Composing a musical tune with MIDI is as simple as playing each part on the keyboard and programming these parts into the computer. In this way, you can produce the melody and all the harmony. The waveforms of music, no matter how complex, are all converted into digital high and low states. That way, they can be stored on any digital medium or exchanged over the Internet. If you plan to send or receive many MIDI files over the Internet, it helps if you have a high-speed connection.

AUDIO MIXING

Imagine that you have created a composition by synthesizing the sounds of several instruments. You have recorded a separate left and right channel for each instrument. You can listen to all the channels together using an amplifier/speaker combination for each instrument. In order to create a single recording of the whole composition, you must combine the signals at just the levels you want. An *audio mixer* can do this.

Audio mixing involves more than connecting the output terminals from various devices together in parallel. If that were done, there would be unwanted interaction among the devices, and some components might even be damaged. A mixer contains an electronic *isolation circuit* that prevents changes in the impedance or other characteristics of any input device from influencing the behavior of any other input device. The isolation circuit requires careful engineering and, along with the internal equalizer, constitutes the most sophisticated part of an audio mixer.

The audio mixer is an essential component of a recording and reproduction system. It is also known as a *mixing board*, a *mixing console*, or a *sound board*. It can have several channels, each with independent controls such as the following.

- *Trim:* This adjusts the input gain from a microphone or other device. It should be adjusted at a high enough level to provide a reasonably strong, clear signal, but not so high as to produce distortion.
- *Equalizer:* This is a graphic equalizer (in low-end mixers) or parametric equalizer (in high-end mixers) that facilitates adjustment of the frequency-response characteristic of the channel.
- *Auxiliary send:* This control allows some of the signal to be sent to an external device such as an amplifier, monitoring headset, or circuit designed to produce special effects.
- *Pan:* This is the equivalent of the balance control found on stereo amplifiers. It has the effect of audibly positioning the signal to right-of-center, near the center, or to left-of-center along a perceived linear axis.
- *Mute:* This is a switch that silences a selected channel. It does not affect the other channels.
- *Solo:* This is a switch that silences every channel except the one selected.
- *Assign:* This is a switch that determines where the output signal from the selected channel is sent.
- *Channel fader:* This is the equivalent of a volume control connected into the selected channel output. It can be either a rotatable knob or a slider. It affects only the output volume of the selected channel.
- *Master fader:* This is the equivalent of a volume control for the output of the entire mixer. After mixing has taken place, adjusting this control causes the level of the whole signal to increase or decrease.

PROBLEM 13-5

In a technical paper, I recently came across the term *pink noise*. What is this? Is it anything like white noise? How are white noise and pink used in audio engineering?

SOLUTION 13-5

Pink noise is similar to white noise with exaggerated bass and attenuated treble. True white noise sounds as if the high-frequency components are louder than the low-frequency components because the human ear/mind is more sensitive to high audio frequencies than to low audio frequencies. Figure 13-11A is a graph of white noise, showing spectral power density as a function of frequency. Figure 13-11B is a graph of pink noise, in which the spectral power density decreases logarithmically as the frequency increases. White and pink noise are used to test the gain-vs.-

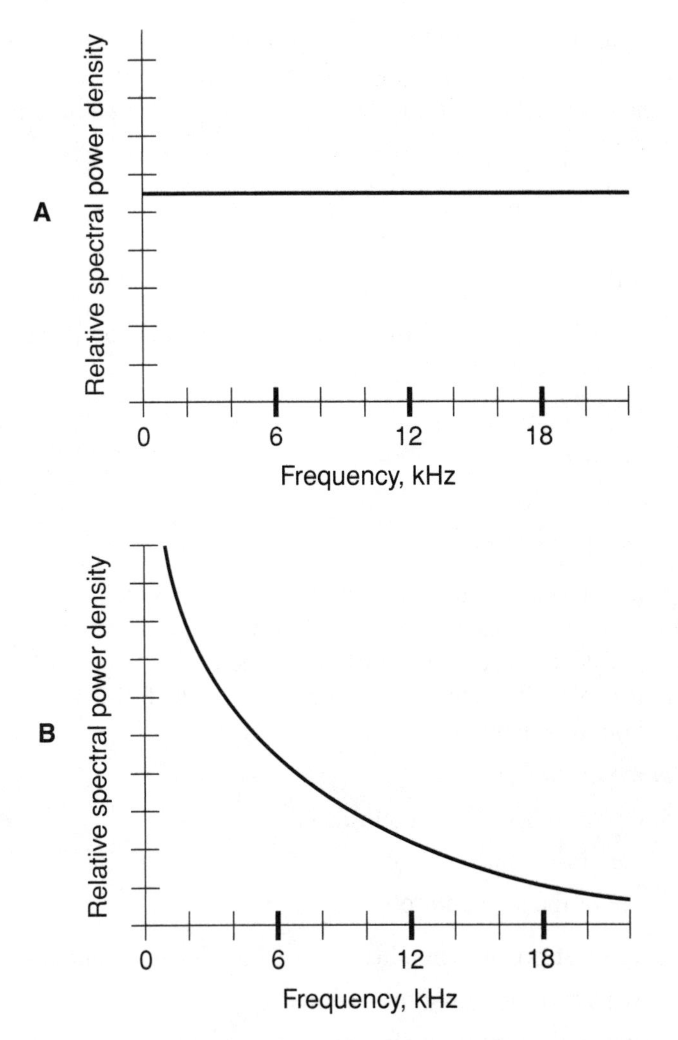

Figure 13-11 At A, white noise has equal spectral power density at all frequencies. At B, pink noise exhibits decreasing spectral power density with increasing frequency.

frequency characteristics of audio systems. White noise is more often employed for quantitative testing (gain measurements) and pink noise is used for qualitative testing (how a system sounds to human listeners).

Quiz

This is an "open book" quiz. You may refer to the text in this chapter. A good score is 8 correct. Answers are in the back of the book.

1. Suppose a source of white noise has a spectral power density of 60 μW/Hz throughout the entire audio spectrum. What, in theory, is the amount of audio power contained in the band between 2.4 kHz and 2.7 kHz?
   (a) 5 μW.
   (b) 18 μW.
   (c) 5 mW.
   (d) 18 mW.

2. The pattern of emphasis on the syllables in a spoken command, declaration, or question are collectively known as
   (a) prosodic features.
   (b) phonemes.
   (c) accents.
   (d) pulse modulation.

3. Suppose you are mixing the signals from 10 channels, each containing a recording of a synthesized signal for a particular musical instrument. At one point, you decide you want to silence three of the channels for a period of eight musical beats. Which, and how many, of the following mixer controls would you most likely use?
   (a) Three of the pan controls.
   (b) Seven of the equalizer controls.
   (c) Three of the mute controls.
   (d) Seven of the solo controls.

4. The number of possible amplitude levels in an analog audio signal is
   (a) known as the sampling resolution.
   (b) known as the sampling frequency.
   (c) usually equal to a power of two.
   (d) theoretically infinite.

5. A plastic medium on which data can be written once by means of a laser (but not overwritten) is called
   (a) vinyl disc.
   (b) CD-R.
   (c) CD-RW.
   (d) DAT.

6. Which of the following media lends itself best to multi-generation copies of voices or music (that is, copies of copies of copies…)?
   (a) Vinyl disc.
   (b) DAT.
   (c) Analog audio tape.
   (d) All of the above work equally well for making multi-generation copies.
7. Allophones in human speech sounds are made up of
   (a) digital signals in ASCII format.
   (b) analog high and low states.
   (c) phonemes with variable characteristics.
   (d) Forget it! Allophones are the most basic components of human speech sounds.
8. Which of the following storage locations can cause physical damage to magnetic tape?
   (a) Inside an enclosed car on a warm, sunny day.
   (b) On a shelf next to books and other papers.
   (c) In close proximity with CD-R or CD-RW media.
   (d) All of the above
9. A comb generator is specifically designed to produce
   (a) an audio signal at specified fundamental frequency only.
   (b) audio signals at a specified fundamental frequency and its harmonics.
   (c) audio pink noise.
   (d) sculpted audio white noise.
10. The sampling rate in A/D conversion should normally be
    (a) at least 8 kHz.
    (b) at least 44.1 kHz.
    (c) at least twice the highest frequency in the analog signal.
    (d) equal to some power of two.

Final Exam

Do not refer to the text when taking this exam. A good score is at least 75 correct. Answers are in the back of the book. It's best to have a friend check your score the first time, so you won't memorize the answers if you want to take the exam again.

1. In a common-base JFET amplifier fed by a sine wave audio signal having a constant frequency, the output signal
   (a) is in phase with the input signal.
   (b) is 90º of phase ahead of the input signal.
   (c) is 90º of phase behind the input signal.
   (d) is inverted with respect to the input signal.
   (e) varies in phase with respect to the input signal.

2. In speaker technology, the term *hangover* refers to
   (a) the phase difference between the applied AF wave and the actual sound wave produced in the surrounding air.
   (b) the extent to which a speaker cone extends beyond the edges of the frame.
   (c) the length of time it takes for a speaker cone to stop moving after an intense audio peak has passed.
   (d) the natural tendency of a speaker to be most sensitive at the low frequencies, and less sensitive at the high frequencies.
   (e) The extent to which a crossover network improperly allows bass AF energy to the tweeter, and/or treble AF energy to the woofer.

3. What is the period of an acoustic sine wave with a frequency of 15 kHz?
   (a) 6.7 μs.
   (b) 67 μs.
   (c) 670 μs.
   (d) 6.7 ms.
   (e) More information is necessary to answer this question.

4. An acoustic sine wave at 15 kHz has a frequency
   (a) below the human hearing range.
   (b) within, but near the low end of, the human hearing range.
   (c) in the middle of the human hearing range.
   (d) within, but near the high end of, the human hearing range.
   (e) above the human hearing range.

5. If two complex impedances, $R_1 + jX_1$ and $R_2 + jX_2$, are perfectly matched and are both purely resistive, then which of the following *must* be true?
   (a) $R_1 = 0$ and $R_2 = 0$.
   (b) $R_1 = X_2$ and $X_1 = R_2$.
   (c) $X_1 = 0$ and $X_2 = 0$.
   (d) $R_1 = X_1$ and $R_2 = X_2$.
   (e) None of the above

6. One of the most serious problems with old-fashioned mercury-vapor voltage regulator tubes is the fact that
   (a) the filaments tend to wear out quickly because the mercury corrodes them.
   (b) the UV emission they produce can be harmful to personnel in their vicinity.
   (c) the mercury can damage the environment when the tube is improperly discarded.
   (d) they are electrically fragile and are more susceptible than semiconductor voltage regulators to damage from transients.
   (e) they only work at voltages too low for most electronic equipment.

7. Suppose an NPN bipolar transistor, connected in a common-emitter configuration, has a dynamic current amplification factor of 100. A certain change in the instantaneous base current, call it dI_B, causes the instantaneous collector current to change by 125 μA. What is dI_B?

(a) It is impossible to calculate this with the information given here.
(b) 12.5 mA.
(c) 1.25 mA.
(d) 12.5 μA.
(e) 1.25 μA.

8. Figure Exam-1 shows a low-power stereo tone control system with two independent adjustments for each channel. What would happen if capacitor X were to fail, resulting in an open circuit?
   (a) The potentiometer in parallel with capacitor X would act as a left-channel treble-attenuation control, rather than as a left-channel bass-attenuation control.
   (b) The potentiometer in parallel with capacitor X would act as a left-channel volume control, rather than as a left-channel treble-attenuation control.
   (c) The potentiometer in parallel with capacitor X would act as a left-channel volume control, rather than as a left-channel bass-attenuation control.
   (d) The potentiometer in parallel with capacitor X would act as a left-channel bass-attenuation control, rather than as a left-channel treble-attenuation control.
   (e) There would be no output from the headset in the left channel.

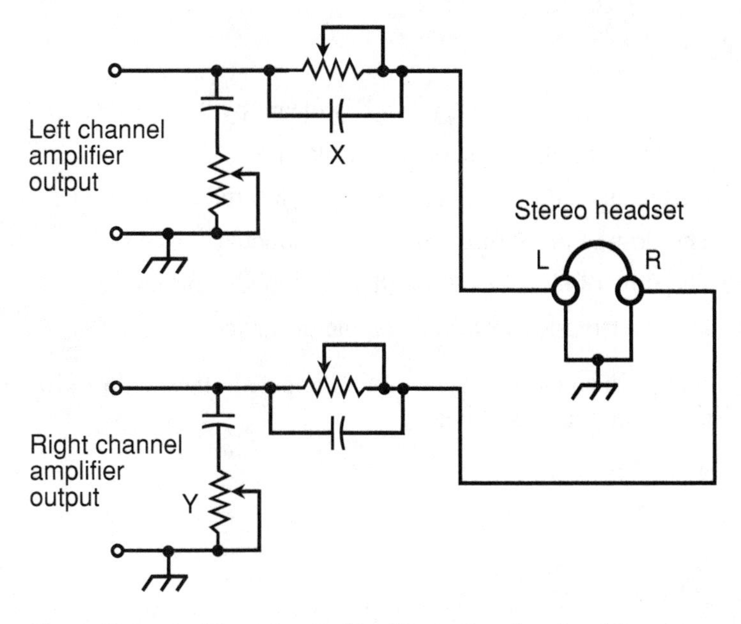

Figure Exam-1 Illustration for Final Exam Questions 8 and 9.

9. In the circuit of Fig. Exam-1, what would happen if potentiometer Y were to burn out, resulting in an open circuit?
   (a) There would be no output from the headset in the right channel.
   (b) The capacitor in series with Potentiometer Y would immediately burn out also.
   (c) It would be impossible to adjust the bass attenuation in the right channel.
   (d) It would be impossible to adjust the treble attenuation in the right channel.
   (e) It would be impossible to adjust either the bass attenuation or the treble attenuation in the right channel.

10. Consider a matrix of 12 resistors, all having identical resistances, and each rated at 2 W. There are four sets of three series-connected resistors, and these four sets are connected in parallel with each other. What is the total power-handling capacity of this network of resistors?
    (a) We must know the actual resistance of each component in order to answer this.
    (b) 24 W.
    (c) 12 W.
    (d) 8 W.
    (e) 6 W.

11. In semiconductor technology, what is meant by the term *doping*?
    (a) Overheating of a semiconductor material.
    (b) Deliberate addition of an impurity to a semiconductor element.
    (c) The flow of holes through a semiconductor device.
    (d) The flow of electrons through a semiconductor device.
    (e) Improper biasing of a semiconductor device.

12. If a change in sound volume is not anticipated, the smallest difference that a listener can notice is about
    (a) ± 0.5 dB.
    (b) ± 1 dB.
    (c) ± 3 dB.
    (d) ± 10 dB.
    (e) ± 20 dB.

13. In a dynamic loudspeaker, the mechanical motion of the diaphragm is produced by the force between a coil, to which the AF signal is applied, and
    (a) a capacitor that alternately charges up with, and then discharges, AF energy.
    (b) an AC electromagnet.
    (c) a permanent magnet or DC electromagnet.
    (d) another coil placed at right angles to it.
    (e) the metal frame of the speaker cone.

14. Consider an attenuator pad made up of resistors. Suppose it is designed to produce a 20-dB power loss at all audio frequencies. If the attenuator input terminals are connected to the output of an amplifier delivering 30 W rms of audio power, what is the audio power available at the attenuator output terminals?
    (a) More information is needed to answer this.
    (b) 10 W rms.
    (c) 3 W rms.
    (d) 300 mW rms.
    (e) 100 mW rms.

15. Suppose the audio signal power at the speaker terminals of a receiver is 10 W, and the noise level is 5 mW. Then the signal-to-noise ratio is
    (a) 3 dB.
    (b) 6 dB.
    (c) 33 dB.
    (d) 66 dB.
    (e) 2000 dB.

16. Vacuum tubes typically operate with DC power-supply plate voltages ranging from about
    (a) 0.5 V to 1.5 V.
    (b) 1.5 V to 12 V.
    (c) 12 V to 50 V.
    (d) 50 V to several kilovolts.
    (e) 0.5 V to several megavolts.

17. When all other factors are held constant, the ripple output from an AC power supply tends to increase as the
    (a) voltage output increases.
    (b) current demand increases.
    (c) load resistance increases.
    (d) load impedance increases.
    (e) load reactance increases.

18. The time-domain graph shown in Fig. Exam-2 represents a signal wave of the sort that contains
    (a) energy at the fundamental frequency, and also at the second harmonic.
    (b) energy at the fundamental frequency, and also at the second and third harmonics.
    (c) energy at the fundamental frequency, and also at the second, third, and fourth harmonics.
    (d) energy at the fundamental frequency and all possible harmonics.
    (e) energy at the fundamental frequency only, and no energy at any harmonics.

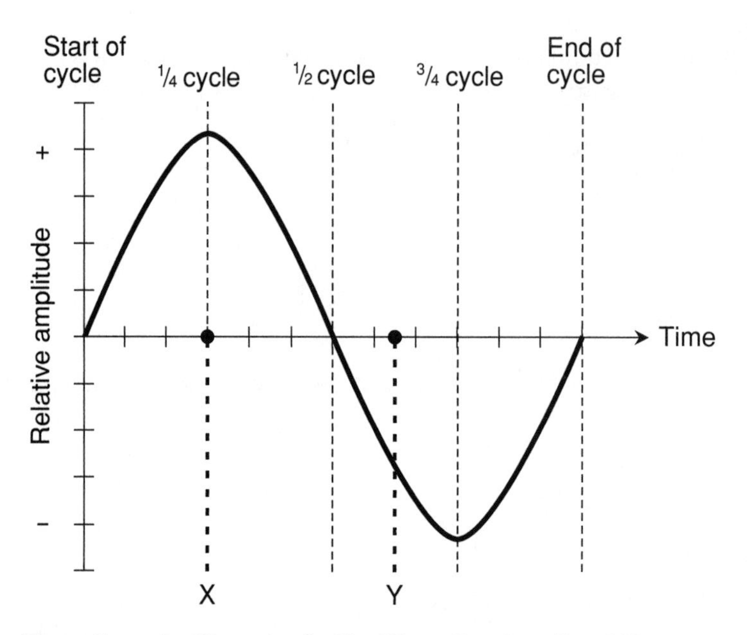

Figure Exam-2 Illustration for Final Exam Questions 18 and 19.

19. In Fig. Exam-2, what is the difference in phase between the points marked X and Y with respect to the waveform shown?
    (a) 0°.
    (b) 45°.
    (c) 90°.
    (d) 135°.
    (e) In order to answer this, we must know the frequency.

20. The cathode in a vacuum tube is analogous to
    (a) the gate in a P-channel JFET.
    (b) the drain in an N-channel MOSFET.
    (c) the collector in a PNP bipolar transistor.
    (d) the base in an NPN bipolar transistor.
    (e) the source in an N-channel JFET.

21. Placing an electronic circuit in a bath of liquid helium represents a radical, but effective, way to almost completely eliminate
    (a) electromagnetic interference.
    (b) thermal noise.
    (c) nonlinearity.
    (d) overdrive.
    (e) feedback.

22. The instantaneous rate of change in a variable voltage is defined as
    (a) the rate at which the voltage increases or decreases at a specific point in time.
    (b) the average rate at which the voltage increases or decreases over a defined period of time.
    (c) the number of times per second that the a signal jumps from one voltage to another instantaneously.
    (d) the difference between the positive peak voltage and the negative peak voltage of an AC wave at a specific point in time.
    (e) the peak-to-peak voltage of an AC wave at a specific point in time.

23. Which of the following effects in an audio system can be attributed to microphonics?
    (a) Nonlinearity of the amplifiers driving the speaker system.
    (b) Excessive negative feedback, resulting in instability.

(c) Excessive susceptibility of the amplifiers to electromagnetic interference.
(d) Unwanted modulation of the oscillators in the tuner.
(e) All of the above

24. In order for a power-line transient suppressor, also called a "surge suppressor" or "surge protector," to work properly, which of the following is necessary?
    (a) The installation of an RF choke in series with the live power line.
    (b) The installation of a capacitor between the live power line and ground.
    (c) The use of a properly grounded three-wire electrical system.
    (d) The use of shielded power cable.
    (e) All of the above

25. If a component contains a pure negative reactance, that component is most likely
    (a) a resistor.
    (b) an inductor.
    (c) a microphone.
    (d) a speaker.
    (e) a capacitor.

26. A monaural headset
    (a) always has a flat frequency response.
    (b) has excellent channel separation.
    (c) is generally used for communications, not for stereo hi-fi listening.
    (d) can respond to only one audio frequency.
    (e) has only one earphone.

27. Fill in the blank in the following sentence to make it true: "In the waveform at the output of an audio amplifier to which a pure sawtooth-wave input signal of constant pitch is applied, the function of ________ repeats itself in identical fashion over and over."
    (a) instantaneous frequency vs. time
    (b) instantaneous voltage vs. time
    (c) instantaneous impedance vs. time
    (d) instantaneous resistance vs. time
    (e) instantaneous conductance vs. time

28. Consider an audio sine wave with a frequency of 440 Hz. The second *derivative* of this has a frequency of

    (a) 110 Hz.

    (b) 220 Hz.

    (c) 440 Hz.

    (d) 880 Hz.

    (e) 1760 Hz.

29. Figure Exam-3 shows a transformer for matching the output of an audio amplifier to a high-impedance headset. Suppose that the number of turns in the primary winding, T_{pri}, is equal to 200. Further suppose that the output impedance of the amplifier is 60 Ω nonreactive and the impedance of the headset is 600 Ω nonreactive. What is the required value for T_{sec}, the number of turns in the secondary winding of the transformer, to ensure a perfect impedance match between the amplifier and the headset?

    (a) More information is needed to answer this.

    (b) T_{sec} = 20.

    (c) T_{sec} = 63.

    (d) T_{sec} = 632.

    (e) T_{sec} = 2000.

30. Fill in the blank in the following statement to make it true: "The extent of the ________ that a component produces is roughly proportional to the current that passes through it."

    (a) shot noise

    (b) reactance

    (c) electromagnetic interference

    (d) audio input power

    (e) attenuation

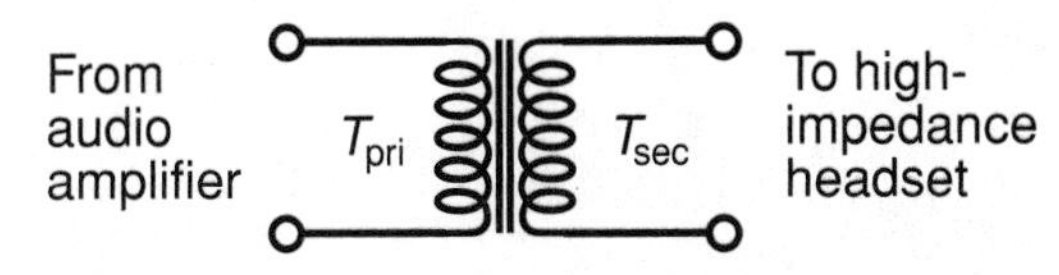

Figure Exam-3 Illustration for Final Exam Question 29.

31. In a PNP bipolar transistor, the N-type semiconductor constitutes the
    (a) drain.
    (b) gate.
    (c) cathode.
    (d) emitter.
    (e) base.

32. A good hi-fi audio amplifier has reasonably constant gain over a range of frequencies spanning at least
    (a) 3 orders of magnitude.
    (b) 10 orders of magnitude.
    (c) 20 orders of magnitude.
    (d) 30 orders of magnitude.
    (e) 40 orders of magnitude.

33. In a P-channel JFET, the majority charge carriers are
    (a) electrons.
    (b) protons.
    (c) neutrons.
    (d) holes.
    (e) positrons.

34. If n resistors of equal value (where n is a positive whole number) are connected in series, the total resistance is equal to
    (a) $1/n^2$ of the resistance of any one of the resistors alone.
    (b) $1/n$ of the resistance of any one of the resistors alone.
    (c) the resistance of any one of the resistors alone.
    (d) n times the resistance of any one of the resistors alone.
    (e) n^2 times the resistance of any one of the resistors alone.

35. The extent to which the output waveform of an amplifier is a faithful reproduction of the input waveform is called
    (a) alpha cutoff.
    (b) beta.
    (c) gamma.
    (d) rolloff.
    (e) linearity.

36. Which of the following statements (a), (b), (c), or (d), if any, is false?
    (a) Sferics are caused mainly by lightning.
    (b) Sferics propagate farther on the AM broadcast band than on the FM broadcast band.
    (c) Sferics originate mainly inside electronic components.
    (d) Sferics sometimes sound like bursts of "static."
    (e) All of the above statements (a), (b), (c), and (d) are true.

37. Figure Exam-4 is supposed to be a schematic diagram of a generic audio amplifier circuit using an N-channel JFET. What, if anything, is wrong with this circuit?
    (a) The drain isn't getting any DC supply voltage.
    (b) The output is short-circuited.
    (c) The input is short-circuited.
    (d) The polarity of the power supply is wrong.
    (e) Nothing is wrong with it.

38. Which of the following statements concerning graphic equalizers is false?
    (a) Active devices, if any, should introduce a minimum of electrical noise.
    (b) Each individual gain control should operate independently of the others.
    (c) The filter frequencies should be carefully chosen and spaced.
    (d) The filters should not affect the signal waveforms, but only their amplitudes.
    (e) The left and right channel outputs should be connected together.

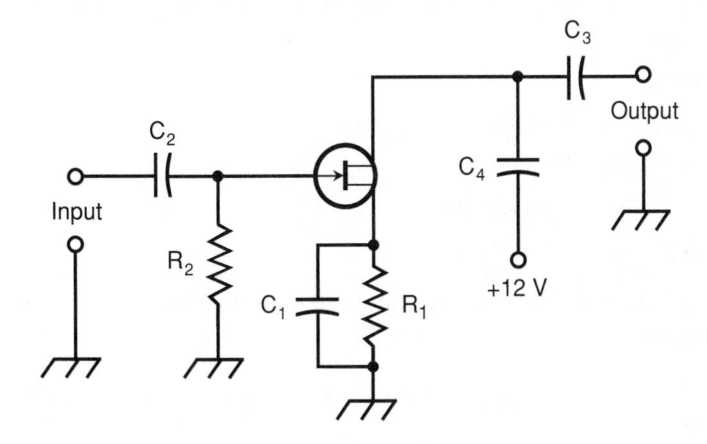

Figure Exam-4 Illustration for Final Exam Question 37.

39. Fill in the blank in the following statement to make it true: "Any NPN bipolar transistor, connected in a common-emitter circuit in which the bias, power-supply voltage, and AC input signal amplitude remain constant, will exhibit ________ that decreases as the input signal frequency increases."
    (a) current loss
    (b) voltage attenuation
    (c) negative feedback
    (d) current gain
    (e) output impedance

40. In an expression of complex impedance $R + jX$, the value of R can never be
    (a) larger than X.
    (b) smaller than X.
    (c) greater than 0.
    (d) less than 0.
    (e) less than the square root of X.

41. At the negative peak of an AC sine wave, the instantaneous rate of amplitude change is
    (a) zero.
    (b) positive and increasing.
    (c) positive and decreasing.
    (d) negative and increasing.
    (e) negative and decreasing.

42. At the positive peak of an AC sine wave, the instantaneous rate of amplitude change is
    (a) zero.
    (b) positive and increasing.
    (c) positive and decreasing.
    (d) negative and increasing.
    (e) negative and decreasing.

43. An attenuator can be used in the input circuit of an audio power amplifier to prevent
    (a) linearity.
    (b) cutoff.
    (c) pinchoff.

(d) saturation.
(e) overdrive.

44. Which of the following terms represents an expression of the extent to which a speaker emits acoustic waves in all directions, as opposed to focusing the waves in one direction?
(a) Rolloff.
(b) Dispersion.
(c) Damping.
(d) Feedback.
(e) Attenuation.

45. Suppose a 12-V battery is connected directly to a 240-Ω resistor. What is the current through the resistor?
(a) 2880 A.
(b) 20 A.
(c) 2.88 A.
(d) 50 mA.
(e) 2.88 mA.

46. When an electromagnetic CRT is used to analyze the waveform of an audio signal, the vertical deflection coils usually receive
(a) the waveform to be analyzed.
(b) a sine-wave signal for synchronization purposes.
(c) a sine-wave signal for frequency adjustment purposes.
(d) a sawtooth wave for horizontal sweep.
(e) no input at all.

47. A source-follower JFET circuit can act as a
(a) parametric equalizer.
(b) high-gain audio amplifier.
(c) sensitive audio oscillator.
(d) broadband impedance step-down transformer.
(e) tone control.

48. Suppose that the four vector drawings in Fig. Exam-5 represent two AC sine waves having identical frequency, but in various phase relationships. The radial distance from the center of the graph is directly proportional to instantaneous amplitude. Which drawing shows wave *X* leading wave *Y* by 90°?

(a) Drawing A.
(b) Drawing B.
(c) Drawing C.
(d) Drawing D.
(e) None of them.

49. Which drawing in Fig. Exam-5 shows waves *X* and *Y* in phase opposition?
(a) Drawing A.
(b) Drawing B.
(c) Drawing C.
(d) Drawing D.
(e) None of them

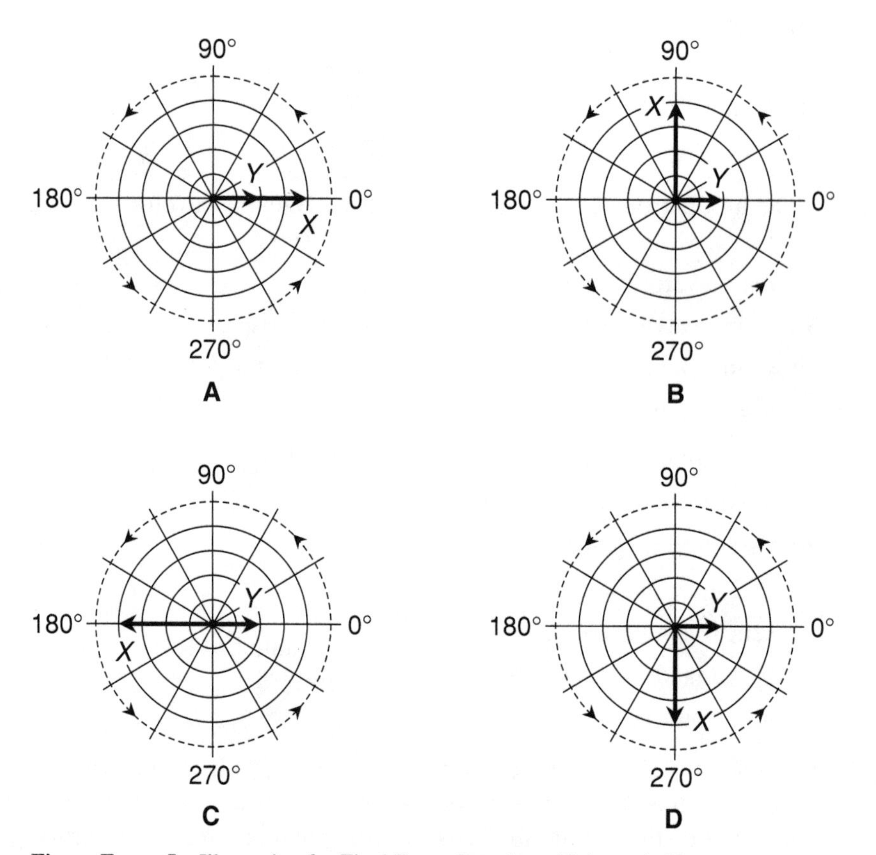

Figure Exam-5 Illustration for Final Exam Questions 48 through 50.

50. In Fig. Exam-5, how do the positive and negative peak amplitudes of waves *X* and *Y* compare, assuming neither wave has a DC component?
    (a) The positive and negative peak amplitudes of *X* are the same as the positive and negative peak amplitudes of *Y*.
    (b) The positive and negative peak amplitudes of *X* are twice the positive and negative peak amplitudes of *Y*.
    (c) The positive and negative peak amplitudes of *X* are half the positive and negative peak amplitudes of *Y*.
    (d) It depends on the frequency.
    (e) It is impossible to infer that information from these drawings.

51. Fill in the blank to make the following statement true: "The voltage that a DC source would exhibit in order to have the same effective voltage as a given AC wave is known as the ________ voltage of the AC wave."
    (a) root-mean-square
    (b) average
    (c) peak
    (d) peak-to-peak
    (e) instantaneous

52. A vacuum-tube amplifier requires some time to "warm up" from a cold start because
    (a) the filaments need time to reach the proper operating temperature.
    (b) the power supply needs time to reach the high voltage required for operating vacuum tubes.
    (c) the capacitors in the circuit need time to charge up.
    (d) the elemental material inside the tubes needs time to condense.
    (e) the microcomputers need time to initialize.

53. The primary-to-secondary voltage-transfer ratio of an audio transformer, assuming source and load impedances that contain no reactance, is equal to
    (a) the square root of the primary-to-secondary impedance-transfer ratio.
    (b) the primary-to-secondary impedance-transfer ratio.
    (c) the square of the primary-to-secondary impedance-transfer ratio.
    (d) the secondary-to-primary impedance-transfer ratio.
    (e) None of the above

54. The primary-to-secondary turns ratio of an audio transformer, assuming source and load impedances that contain no reactance, is equal to
    (a) the square root of the primary-to-secondary impedance-transfer ratio.
    (b) the primary-to-secondary impedance-transfer ratio.
    (c) the square of the primary-to-secondary impedance-transfer ratio.
    (d) the secondary-to-primary impedance-transfer ratio.
    (e) None of the above

55. In a common-source MOSFET amplifier fed by a sine wave audio signal having a constant frequency, the output signal
    (a) is in phase with the input signal.
    (b) is 90° of phase ahead of the input signal.
    (c) is 90° of phase behind the input signal.
    (d) is inverted with respect to the input signal.
    (e) varies in phase with respect to the input signal.

56. An acoustic disturbance in the atmosphere is
    (a) a transverse wave.
    (b) a longitudinal wave.
    (c) an elliptical wave.
    (d) a sine wave.
    (e) an irregular wave.

57. The graph of Fig. Exam-6 shows I_D as a function of E_G for a hypothetical N-channel JFET. For weak-signal audio amplification with good linearity, optimum DC gate bias voltage is indicated by
    (a) point *X*.
    (b) point *Y*.
    (c) point *Z*.
    (d) the range from point *X* to point *Y*.
    (e) the range from point *Y* to point *Z*.

58. Which of the points or ranges in Fig. Exam-6 represents DC gate bias that places the device in a state of saturation under no-signal conditions?
    (a) point *X*.
    (b) point *Y*.

(c) point *Z*.
(d) the range from point *X* to point *Y*.
(e) the range from point *Y* to point *Z*.

59. Which of the points or ranges in Fig. Exam-6 represents DC gate bias that places the device in a state of pinchoff under no-signal conditions?
(a) point *X*.
(b) point *Y*.
(c) point *Z*.
(d) the range from point *X* to point *Y*.
(e) the range from point *Y* to point *Z*.

60. A theoretically perfect op amp would have
(a) zero output impedance at all frequencies.
(b) infinite output impedance at all frequencies.
(c) equal input and output impedances at all frequencies.
(d) purely inductive input impedance at all frequencies.
(e) purely capacitive input impedance at all frequencies.

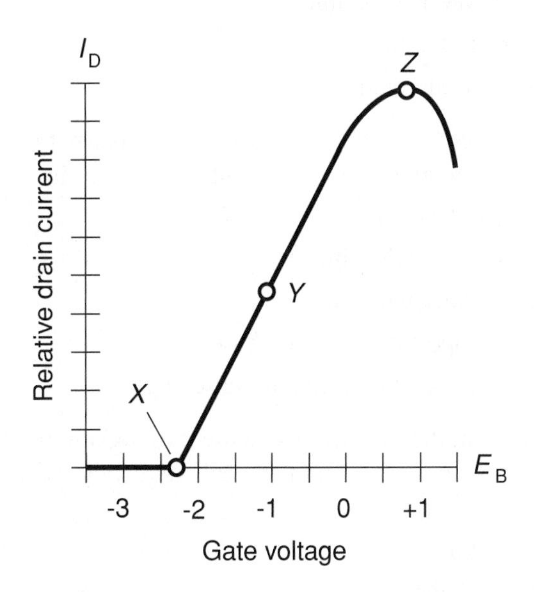

Figure Exam-6 Illustration for Final Exam Questions 57 through 59.

61. An audio gain of +16 dB represents an increase in sound power by a factor of
    (a) 4.
    (b) 16.
    (c) 40.
    (d) 256.
    (e) None of the above

62. What is the fourth harmonic of an AC wave with a peak-to-peak amplitude of 16 V?
    (a) 4 V.
    (b) 16 V.
    (c) 64 V.
    (d) It depends on whether or not the signal has a DC component.
    (e) In order to answer this, we must know the fundamental frequency.

63. The maximum extent, in decibels, to which an audio filter can block signals at unwanted frequencies far removed from desired frequencies is known as the
    (a) zero-point power attenuation.
    (b) half-point power attenuation.
    (c) ultimate power attenuation.
    (d) cutoff power attenuation.
    (e) rolloff power attenuation.

64. In the rotating-vector model of an AC sine wave with no DC component having constant frequency and constant amplitude, the length of the vector
    (a) becomes greater with time.
    (b) becomes smaller with time.
    (c) remains constant with time.
    (d) is directly proportional to the phase angle.
    (e) is inversely proportional to the phase angle.

65. Fill in the blank in the following statement to make it true: "In a ________, the gain is unity above a certain critical frequency. Below that frequency, the gain is adjustable."
    (a) bandpass filter
    (b) graphic equalizer
    (c) treble level control

(d) low-frequency shelf filter
(e) notch filter

66. Which of the following can be a source of electromagnetic interference (EMI) to hi-fi equipment?
    (a) A nearby incandescent light bulb.
    (b) A nearby personal computer.
    (c) A low-impedance microphone.
    (d) Excessive linearity in an amplifier.
    (e) All of the above

67. Which characteristic differs between Filter X and Filter Y, as evidenced by comparing the attenuation-vs.-frequency curves in Fig. Exam-7?
    (a) The skirt slope.
    (b) The half-power point.
    (c) The ultimate attenuation.
    (d) The minimum attenuation.
    (e) None of the above

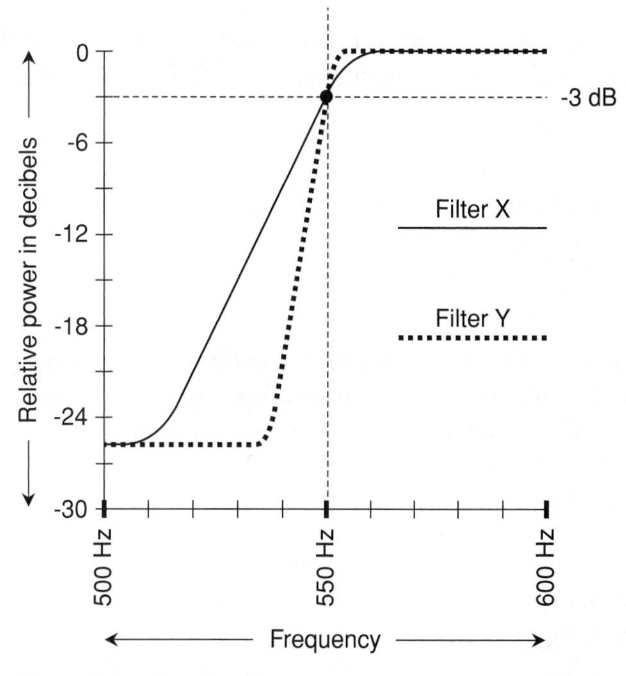

Figure Exam-7 Illustration for Final Exam Question 67.

68. Suppose an FM tuner is set for 101.7 MHz. An interfering signal is heard, and you suspect that it is a harmonic from a radio transmitter somewhere in the vicinity. If this is in fact the case, on which of the following output frequencies might the offending transmitter be operating?
    (a) 305.1 MHz.
    (b) 203.4 MHz.
    (c) 33.9 MHz.
    (d) 10.1 MHz.
    (e) Any of the above

69. The difference, in decibels, between the strongest and the weakest audio output signals that a hi-fi system can produce while maintaining good linearity is known as the
    (a) dynamic range.
    (b) alpha cutoff.
    (c) beta.
    (d) distortion coefficient.
    (e) gain bandwidth product.

70. Which of the following circuits or components is best for impedance matching between the output of an audio hi-fi amplifier and a speaker system?
    (a) An *LC* network.
    (b) A crossover network.
    (c) A lowpass filter.
    (d) A highpass filter.
    (e) A transformer.

71. Consider an audio power-amplifier circuit with a triode vacuum tube. Suppose the cathode is at DC ground, and the plate power-supply voltage is +450 V DC. Which of the following represents a likely control-grid power-supply voltage?
    (a) +450 V DC.
    (b) +100 V DC.
    (c) −100 V DC.
    (d) −450 V DC.
    (e) −900 V DC.

72. Which of the following conditions (a), (b), (c), or (d), if any, cannot be tolerated in a good high-fidelity audio amplifier?
    (a) Low gain.
    (b) Low input impedance.
    (c) Nonlinearity.
    (d) Negative feedback.
    (e) All of the above can be tolerated in a good high-fidelity audio amplifier.

73. In an analog audio tape recorder operating in the normal record mode, the tape moves past an erase head before anything is recorded on it. This prevents
    (a) overdrive.
    (b) harmonic generation.
    (c) distortion.
    (d) doubling.
    (e) echo effect.

74. In a bipolar transistor, a small change in I_B causes a large variation in I_C when the bias is just right. This makes it possible for the device to function as
    (a) a voltage limiter.
    (b) a tone control.
    (c) an attenuator.
    (d) a current amplifier.
    (e) a parametric equalizer.

75. Consider an audio power-amplifier circuit with a pentode vacuum tube. Suppose the cathode is at DC ground, and the plate power-supply voltage is +600 V DC. Which of the following represents a likely screen-grid power-supply voltage?
    (a) +600 V DC.
    (b) +170 V DC.
    (c) 0 V DC.
    (d) –200 V DC.
    (e) –600 V DC.

76. The efficiency of a well-designed, properly operating class-B push-pull audio power amplifier is usually
    (a) over 65 percent.
    (b) 50 to 65 percent.
    (c) 25 to 50 percent.
    (d) less than 25 percent.
    (e) the same as a class-A amplifier with the same DC power input.
77. Consider the DC circuit shown in Fig. Exam-8. What is the resistance *R*?
    (a) 0.35 mΩ.
    (b) 1.4 Ω.
    (c) 28 Ω.
    (d) 35 Ω.
    (e) More information is necessary to answer this.
78. In the circuit of Fig. Exam-8, what is the power dissipated in the potentiometer?
    (a) 0.035 mW.
    (b) 1.4 W.
    (c) 28 W.
    (d) 35 W.
    (e) More information is necessary to answer this.

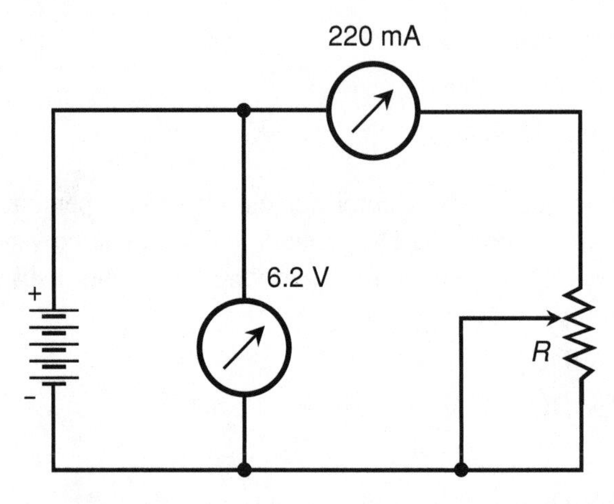

Figure Exam-8 Illustration for Final Exam Questions 77 and 78.

79. An audio filter that produces power gain outside of a defined frequency band, and less gain (or even a loss) within that band, is known as a

    (a) bandpass filter.

    (b) highpass filter.

    (c) rolloff filter.

    (d) parametric filter.

    (e) band-rejection filter.

80. Which of the following factors (a), (b), (c), or (d), if any, does *not* differ significantly between a well-designed home stereo system and a well-designed portable FM stereo headphone radio?

    (a) The cost of the system.

    (b) The amount of audio output power the system can deliver.

    (c) The amount of electrical power required to operate the system.

    (d) The number of stereo channels in the audio output.

    (e) All of the above factors (a), (b), (c), and (d)differ significantly between a well-designed home stereo system and a well-designed portable FM stereo headphone radio.

81. Which of the following represents a typical sampling rate for digital hi-fi music recording and reproduction?

    (a) 330 samples per second

    (b) 3300 samples per second.

    (c) 8800 samples per second.

    (d) 22,500 samples per second.

    (e) 44,100 samples per second.

82. A twin-T oscillator does not produce a perfect sine-wave AC output. This can be a good thing because

    (a) it can help prevent "ear fatigue" that sine-wave audio notes cause.

    (b) it prevents excessive harmonic generation.

    (c) it prevents instability in the feedback loop.

    (d) it prevents overdrive.

    (e) Forget it! It is a bad thing.

83. The base-collector (B–C) junction of a bipolar transistor is normally in a state of
    (a) forward breakover.
    (b) saturation.
    (c) reverse bias.
    (d) zero bias.
    (e) avalanche.

84. If n resistors of equal value (where n is a positive whole number) are connected in parallel, the *net conductance* of the whole set is equal to
    (a) $1/n^2$ of the conductance of any one of the resistors alone.
    (b) $1/n$ of the conductance of any one of the resistors alone.
    (c) the conductance of any one of the resistors alone.
    (d) n times the conductance of any one of the resistors alone.
    (e) n^2 times the conductance of any one of the resistors alone.

85. In an audio mixer, the solo switch
    (a) silences every harmonic except the one selected.
    (b) silences every frequency except the one selected.
    (c) silences every channel except the one selected.
    (d) silences every singing voice except the one selected.
    (e) silences every musical instrument except the one selected.

86. Figure Exam-9 is a schematic diagram of a circuit intended to filter out the ripple from an AC power supply. What, if anything, is wrong with this circuit, and what can be done to correct the problem?
    (a) Nothing is wrong with this circuit. No correction is necessary.
    (b) The polarity of the capacitor is incorrect; it should be reversed.
    (c) The choke should be connected between the positive and negative output terminals, rather than in series.
    (d) The capacitor should be connected between the positive and negative input terminals, rather than in series.
    (e) The choke should be on the input side of the capacitor, rather than on the output side.

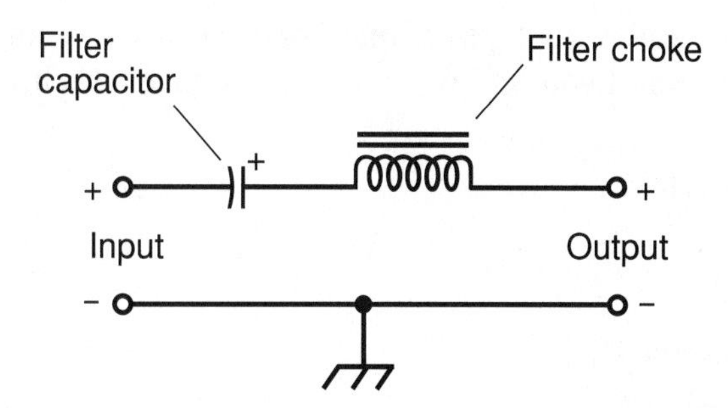

Figure Exam-9 Illustration for Final Exam Question 86.

87. For the purposes of speech recognition or voice communication, an analog signal can be restricted to a band of frequencies as narrow as approximately
    (a) 30 Hz to 300 kHz.
    (b) 30 Hz to 30 kHz.
    (c) 3 kHz to 300 kHz.
    (d) 300 Hz to 30 kHz.
    (e) 300 Hz to 3 kHz.

88. Which of the following transducer types employs a ceramic or quartz crystal and is especially well suited to the transmission of ultrasound?
    (a) Piezoelectric.
    (b) Electrostatic.
    (c) Dynamic.
    (d) Whizzer.
    (e) Woofer.

89. An audio filter that theoretically blocks energy at one frequency while passing energy without attenuation at all other frequencies is a
    (a) notch filter.
    (b) bandpass filter.
    (c) lowpass filter.
    (d) highpass filter.
    (e) midrange filter.

90. Which of the following types of microphone takes advantage of the relative motion between a coil and a magnet to produce AF currents from sound waves?
    (a) Piezoelectric.
    (b) Acoustic.
    (c) Ceramic.
    (d) Electrostatic.
    (e) Dynamic.
91. In the English language there are 40 elementary sounds known as phonemes. Variations in the way these phonemes are uttered (such as pitch, loudness, and duration) produces up to 128
    (a) accents.
    (b) dialects.
    (c) allophones.
    (d) inflections.
    (e) insinuations.
92. Suppose a 220-Ω resistor carries 50 mA of direct current. What is the DC voltage across this resistor?
    (a) 11 V.
    (b) 44 V.
    (c) 1.1 kV.
    (d) 4.4 kV.
    (e) More information is needed to determine this.
93. Which of the following is a good estimate of the highest acoustic-wave frequency that an elderly person is likely to be able to hear?
    (a) 7 Hz.
    (b) 70 Hz.
    (c) 700 Hz.
    (d) 7 kHz.
    (e) 70 kHz.
94. In an amplifier system with automatic gain control (AGC),
    (a) the gain remains constant as the input signal amplitude increases.
    (b) the gain increases as the input signal amplitude increases.
    (c) the gain decreases as the input signal amplitude increases.

(d) the gain increases as the input signal frequency increases.

(e) the gain decreases as the input signal frequency increases.

95. A digital music CD is "played back" by means of

(a) a thin laser beam that is scattered by pits in the disc surface.

(b) an electromagnetic transducer that detects magnetization of the disc surface.

(c) a stylus that passes along a physical groove in the disc surface.

(d) ultrasonic waves that bounce off irregularities in the surface of the disc.

(e) radio waves that pass through more or less "RF-dense" regions in the disc.

96. Which of the following statements (a), (b), (c), or (d), if any, is false?

(a) Digital audio tape provides better fidelity than analog audio tape.

(b) Digital audio tape provides greater dynamic range than analog audio tape.

(c) Digital audio tape provides a better noise figure than analog audio tape.

(d) Digital audio tape records signals that vary continuously, but analog audio tape records signals in discrete levels.

(e) All of the above statements (a), (b), (c), and (d) are true.

97. The graph in Fig. Exam-10 portrays

(a) impulse noise.

(b) pink noise.

(c) white noise.

(d) sferics.

(e) None of the above

98. Suppose an impedance-matching transformer is used to operate a 16-Ω microphone with a preamplifier rated at 576 Ω input impedance. A perfect match, with no reactance in the source or the load, results with the microphone connected to the primary and the amplifier input connected to the secondary. What is the primary-to-secondary voltage-transfer ratio of the transformer under these conditions?

(a) 36/1.

(b) 6/1.

(c) 1/6.

(d) 1/36.

(e) More information is necessary to answer this.

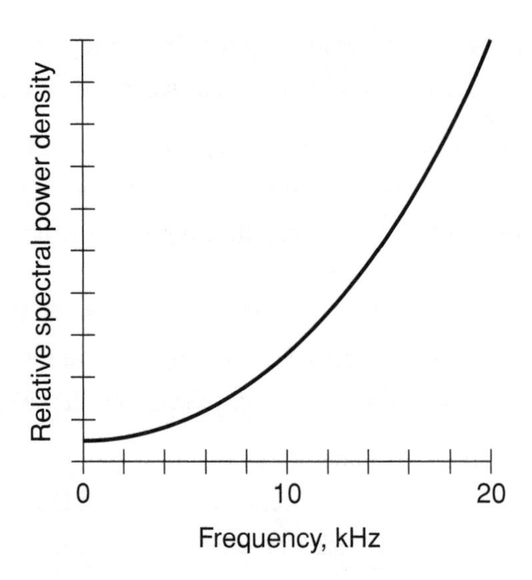

Figure Exam-10 Illustration for Final Exam Question 97.

99. Imagine two audio sine waves, both having a frequency of 850 Hz, and neither having a DC component. Suppose these two sine waves are exactly in phase. If these two waves are combined, the composite wave has a peak-to-peak voltage equal to
    (a) the average of the peak-to-peak voltages of the two waves.
    (b) the difference between the peak-to-peak voltages of the two waves.
    (c) the sum of the peak-to-peak voltages of the two waves.
    (d) twice the sum of the peak-to-peak voltages of the two waves.
    (e) zero.

100. A low-power amplifier should not be operated at or near its maximum output with a speaker system that uses small tweeters because
    (a) the tweeters can produce feedback that causes the amplifier to break into oscillation.
    (b) the resulting high-frequency sound can be harmful to the eardrums.
    (c) the low-frequency output of the amplifier can short out the crossover network and damage the tweeters.
    (d) the amplifier can produce distortion under such conditions, resulting in ultrasonic harmonics with peaks that can damage small tweeters.
    (e) The premise is wrong! It's perfectly all right to use a low-power amplifier at or near its maximum output with small tweeters.

Answers to Quiz Questions

Chapter 1: Direct Current Basics

1. c	2. a	3. d	4. a	5. a
6. b	7. b	8. c	9. a	10. b

Chapter 2: Alternating Current Basics

1. c	2. c	3. a	4. c	5. d
6. d	7. a	8. b	9. c	10. d

Chapter 3: Fundamentals of Phase

1. a	2. b	3. b	4. c	5. a
6. b	7. c	8. c	9. c	10. d

Chapter 4: The Bipolar Transistor

1. a 2. b 3. d 4. b 5. c
6. b 7. d 8. c 9. a 10. b

Chapter 5: The Field-Effect Transistor

1. b 2. c 3. c 4. b 5. d
6. c 7. a 8. d 9. c 10. a

Chapter 6: Electron Tubes

1. c 2. d 3. c 4. d 5. a
6. a 7. b 8. c 9. a 10. a

Chapter 7: Audio Characteristics and Components

1. c 2. b 3. c 4. d 5. a
6. d 7. b 8. d 9. a 10. b

Chapter 8: Speakers, Headsets, and Microphones

1. c 2. a 3. d 4. c 5. a
6. d 7. b 8. b 9. b 10. a

Chapter 9: Impedance-Matching Transformers

1. c 2. d 3. b 4. c 5. d
6. d 7. b 8. a 9. c 10. a

Chapter 10: Transistorized Amplifiers and Oscillators

1. b 2. c 3. b 4. d 5. b
6. a 7. d 8. d 9. b 10. c

Chapter 11: Filters and Equalizers

1. b	2. c	3. d	4. b	5. a
6. d	7. d	8. d	9. b	10. a

Chapter 12: Noise, Hum, Interference, and Grounding

1. d	2. c	3. b	4. b	5. a
6. a	7. c	8. d	9. b	10. c

Chapter 13: Recording, Reproduction, and Synthesis

1. d	2. a	3. c	4. d	5. b
6. b	7. c	8. a	9. b	10. c

Final Exam Answer Key

1. a	2. c	3. b	4. d	5. c
6. c	7. e	8. c	9. d	10. b
11. b	12. c	13. c	14. d	15. c
16. d	17. b	18. e	19. d	20. e
21. b	22. a	23. d	24. c	25. e
26. c	27. b	28. c	29. d	30. a
31. e	32. a	33. d	34. d	35. e
36. c	37. a	38. e	39. d	40. d
41. a	42. a	43. e	44. b	45. d
46. a	47. d	48. b	49. c	50. b
51. a	52. a	53. a	54. a	55. d
56. b	57. b	58. c	59. a	60. a

61. c	62. e	63. c	64. c	65. d
66. b	67. a	68 c	69. a	70. e
71. c	72. c	73. d	74. d	75. b
76. b	77. c	78. b	79. e	80. d
81. e	82. a	83. c	84. d	85. c
86. d	87. e	88. a	89. a	90. e
91. c	92. a	93. d	94. c	95. a
96. d	97. e	98. c	99. c	100. d

Suggested Additional Reading

Gibilisco, S. *Electricity Demystified*. New York, NY: McGraw-Hill, 2005.

Gibilisco, S. *Electronics Demystified*. New York, NY: McGraw-Hill, 2005.

Gibilisco, S. *Teach Yourself Electricity and Electronics, 4th edition*. New York, NY: McGraw-Hill, 2006.

McComb, G. *Electronics for Dummies*. New York, NY: Wiley Press, 2005.

Pittman, Aspen. *The Tube Amp Book*. San Francisco, CA: Backbeat Books, 2003.

Rumreich, Mark. *The Car Stereo Handbook*. New York, NY: McGraw-Hill, 1998.

Slone, G. Randy. *High-Power Audio Amplifier Construction Manual*. New York, NY: McGraw-Hill, 1999.

Slone, G. Randy. *The Audiophile's Project Sourcebook*. New York, NY: TAB/McGraw-Hill, 2002.

Strong, Jeff. *Home Recording for Musicians for Dummies*. New York, NY: Wiley Publishing, Inc., 2002.

Trubitt, Rudy. *Live Sound for Musicians*. Milwaukee, WI: Hal Leonard Corporation, 1997.

White, Ira. *Audio Made Easy*. Milwaukee, WI: Hal Leonard Corporation, 1997.

White, Paul. *Basic* Series. London, England: Sanctuary Publishing, 1999-2003.

Yoder, Andrew. *Auto Audio, Second Edition*. New York, NY: McGraw-Hill, 2000.

INDEX

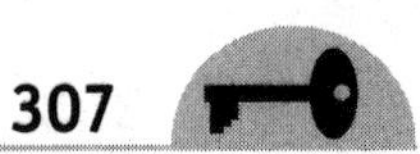